Critical Appraisal of Physical Science as a Human Enterprise

Science & Technology Education Library

VOLUME 36

SCOPE

The book series *Science & Technology Education Library* provides a publication forum for scholarship in science and technology education. It aims to publish innovative books which are at the forefront of the field. Monographs as well as collections of papers will be published.

For other titles published in this series, go to
www.springer.com/series/6512

Mansoor Niaz

Critical Appraisal of Physical Science as a Human Enterprise

Dynamics of Scientific Progress

Springer

Mansoor Niaz
Epistemology of Science Group
Department of Chemistry
Universidad de Oriente
Cumaná, Estado Sucre
Venezuela

ISBN 978-1-4020-9625-9 e-ISBN 978-1-4020-9626-6

Library of Congress Control Number: 2009920028

Printed on acid-free paper

springer.com

For Magda and Sabuhi
For their love, patience, and understanding

Acknowledgments

My institution, Universidad de Oriente (Venezuela) has been the major sponsor of most of my research activities for the last 20 years, through various grants provided by the Consejo de Investigación. I am indebted to Juan Pascual-Leone (York University, Toronto), Richard F. Kitchener (Colorado State University), Michael R. Matthews (University of New South Wales, Australia), Stephen G. Brush (University of Maryland), and Gerald Holton (Harvard University) for helpful discussions on various aspects of history and philosophy of science. A special word of thanks is due to Art Stinner (University of Manitoba, Winnipeg) and colleagues (especially Stephen Klassen, University of Winnipeg) for providing a congenial and stimulating intellectual environment during my three very productive visits in 2001, 2003, and 2007.

Thanks are due to the following publishers for reproduction of materials from my publications: (a) Elsevier (Chapter 5); and (b) Oxford University Press (Chapter 7)

Contents

Chapter 1
Introduction

It is generally believed that doing science means accumulating empirical data with no or little reference to the interpretation of the data based on the scientist's theoretical framework or presuppositions. Holton (1969a) has deplored the widely accepted myth (experimenticism) according to which progress in science is presented as the inexorable result of the pursuit of logically sound conclusions from unambiguous experimental data. Surprisingly, some of the leading scientists themselves (Millikan is a good example) have contributed to perpetuate the myth with respect to modern science being essentially empirical, that is carefully tested experimental facts (free of *a priori* conceptions), leading to inductive generalizations.

Based on the existing knowledge in a field of research a scientist formulates the guiding assumptions (Laudan et al., 1988), presuppositions (Holton, 1978, 1998) and "hard core" (Lakatos, 1970) of the research program that constitutes the imperative of presuppositions, which is not abandoned in the face of anomalous data. Laudan and his group consider the following paraphrase of Kant by Lakatos as an important guideline: philosophy of science without history of science is empty. Starting in the 1960s, this "historical school" has attempted to redraw and replace the positivist or logical empiricist image of science that dominated for the first half of the twentieth century. Among other aspects, one that looms large in these studies is that of "guiding assumptions" and has considerable implications for the main thesis of this monograph (Chapter 2).

Many major steps in science, probably all dramatic changes, and most of the fundamental achievements of what we now take as the advancement or progress of scientific knowledge have been controversial and have involved some dispute or another. Scientific controversies are found throughout the history of science. While nobody would deny that science in the making has had many controversies, most science textbooks and curricula consider it as the uncontroversial rational human endeavor (Machamer et al., 2000).

The objective of this monograph is to reconstruct historical episodes and experiments that have been important in scientific progress, to explore the role played by controversies and rivalries among scientists. Although progress in science has been replete with controversies, scientists themselves either ignore or simply downplay their role. Such presentations lack the appreciation of the dynamics of

M. Niaz, *Critical Appraisal of Physical Science as a Human Enterprise: Dynamics of Scientific Progress*, Science and Technology Education Library 36,

"science-in-the-making." A major problem in scientific progress is the underdetermination of theories by experimental evidence (Duhem–Quine thesis), which leads to the following (among others) consequences: (a) there are no crucial experiments in science, nor instant rationality; (b) empirical evidence does not necessarily refute a theory; and (c) refutation of a theory is a complex and lengthy process that requires the critical appraisal of the scientific community. There has been considerable controversy among philosophers of science on these issues (Chapter 3). For example, according to Popper, confirmations count for little; refutations are what matter. For Lakatos, refutations count for little; confirmations/verifications (heuristic power) are what matter. Duhem (1914) expressed this dilemma in succinct terms: "Experimental verifications are not the base of a theory, but its crown." In other words, in the beginning a scientist does not need to have all the experimental facts to build a theory, and hence facts are not at the base of the theory, but rather its crown, which facilitates verifications (heuristic power).

This monograph provides methodological guidelines based on a historical perspective of the philosophy of science (Duhem to Lakatos) that facilitate an understanding of scientific progress beyond that of inductive generalizations. Furthermore, these guidelines suggest that progress in science is not merely based on accumulation of experimental data, but rather dependent on creative imagination of the scientific community, as can be observed from the following episodes presented in this monograph.

Kinetic theory: Maxwell's presuppositions. Postulates of the kinetic theory were speculative and played the role of simplifying assumptions (Chapter 4). Based on these assumptions, the theorists built a series of tentative models that progressively incorporated the behavior of real gases. Similar to other research programs in the history of science, James C. Maxwell's (1831–1879) was based on inconsistent foundations and had to compete with chemical thermodynamics, a rival research program. Maxwell's (1860) paper was based on "strict mechanical principles" derived from Newtonian mechanics and yet at least two of Maxwell's simplifying assumptions (referring to the movement of particles and the consequent generation of pressure) were in contradiction with Newton's hypothesis explaining the gas laws based on repulsive forces between particles. Later, the work of Boltzmann and van der Waals facilitated a progressive "problemshift" by eliminating some of the simplifying assumptions included in Maxwell's original research program.

Periodic table of the chemical elements: From Mendeleev to Moseley. Dimitri I. Mendeleev's (1834–1907) dilemma was that on the one hand he could rightly claim that the periodic law was based on experimental properties of elements (an aspiration of scientists in the late nineteenth century), and yet he could not give up the bigger challenge, viz., the possible causes of periodicity (Chapter 5). Many students must have wondered how a simple arrangement of the chemical elements could provide predictive and explanatory (accommodation) power to Mendeleev's periodic table? Textbook authors and even many scholars give the impression that for almost 100 years (1820–1920) scientists had no idea or never asked the question as to whether there could be an underlying pattern to explain periodicity. Based on Mendeleev's original writings, I have demonstrated that he did use the atomic theory (long before modern atomic theory was established) to explain periodicity in the periodic table. It is plausible to suggest that based on the work of Dalton, Cannizaro, and others Mendeleev formulated the theoretical framework of his research program that constituted his presuppositions.

Foundations of modern atomic theory: Thomson, Rutherford, and Bohr. J.J. Thomson's (1856–1940) work on cathode rays, E. Rutherford's (1871–1937) work on alpha particles, and N. Bohr's (1885–1962) work on quantum of action laid the foundations of the modern atomic theory (Chapter 6). A reconstruction of the historical events based on original writings of these scientists shows that besides the experiments, it was the controversies and rivalries among these scientists that facilitated an understanding of the experimental data. Scrutinizing Thomson's (1897) article it can be observed that he goes far beyond a simple presentation of experimental results by speculating, hypothesizing, proposing models, offering explanations, and formulating a theory which contrasts with the traditional view of the scientific method. To explain the findings of Rutherford's alpha particle experiments, Thomson was led into a bitter dispute with his student and colleague. To maintain his model of the atom and to explain large angle deflections of alpha particles, Thomson put forward the hypothesis of compound scattering, whereas Rutherford had explained the same experimental finding through his hypothesis of single scattering. This was the major controversy facing atomic models at the beginning of the twentieth century, and the contending parties were led by prominent scientists, who could easily have met over dinner and thrashed out their differences. However, scientific controversies are resistant and require the intervention of the scientific community. Bohr's four postulates were not only based on presuppositions but also on speculations, which according to Lakatos is perfectly justifiable, and helped Bohr to solve the paradox of the stability of the Rutherford atom by postulating a rival research program. Bohr's 1913 model of the atom was again controversial and in general had a fairly adverse reception in the scientific community. For many physicists, Bohr's atom sat like a baroque tower upon the Gothic base of classical electrodynamics. Actually, this illustrates another facet of the dynamics of progress in science, in which scientists resort to contradictory "grafts," a strategy frequently observed in the history of science.

Determination of elementary electrical charge: Millikan and Ehrenhaft. Both Robert Millikan (1868–1953) and Felix Ehrenhaft (1879–1952) used very similar experiments and data (Chapter 7), which led the former to postulate the elementary electric charge (electrons), whereas the latter postulated fractional charges (sub-electrons). The controversy and the rivalry between the two was bitter and lasted for many years (1910–1925). Finally, it was the scientific community that decided in favor of Millikan. Publication of Millikan's handwritten notebooks in 1978 by Holton provided a better understanding of Millikan's procedure. Holton (1978) attributed Millikan's data selection and reduction procedures to his presuppositions with respect to the atomic nature of electricity. Franklin (1981) and Goodstein (2001), however, disagree with Holton's interpretation. Interestingly, however, we have witnessed the controversy not only among the original protagonists (Millikan and Ehrenhaft), but after almost 100 years historians and philosophers of science do not seem to agree as to what really happened. If Millikan's handwritten notebooks had been lost (or for that matter if Holton, Franklin, and Goodstein had not consulted them), the oil drop experiment would have remained an enigma for some, while for others it would have been a classical test case of how experiments unambiguously lead to theoretical formulations. In this chapter, I have presented a critical appraisal of various authors and tried to present a closure to the controversy. A closure to the controversy with respect to the oil drop experiment is possible if we recognize that Millikan's data selection procedure depended primarily on his perseverance with his presuppositions. Hintikka (2006) has added a new dimension to the controversy by arguing that it is a much worse breach of the ethics of science to accuse Millikan of fraud merely because he omitted data.

Paradox of the photoelectric effect: Einstein and Millikan. Einstein's (1879–1955) hypothesis of lightquanta was first presented in 1905 and was not taken seriously by physicists for just over 15 years (Chapter 8). The reason for this was that it implied an unnecessary rejection of the highly verified classical wave theory of radiation. Both Duhem and Lakatos have emphasized that verifications count for more than refutations. Einstein's 1905 theory of lightquanta to explain the photoelectric effect played a crucial role in Millikan's experimental determination of Planck's constant (h) in 1916. Despite this, it is paradoxical that Millikan accepted the experimental data and still rejected Einstein's theory and considered it to be reckless. This may sound incredible to any student of history and philosophy of science. Most physical science textbooks and physicists at present would consider Millikan's (1916b) experimental data as evidence for Einstein's quantum theory (hypothesis of lightquanta). It shows that what we consider these days in retrospect to be strong experimental evidence to support Einstein's theory of lightquanta was not considered to be so by the original protagonists. A historical reconstruction shows that Millikan had a strong presupposition (wave theory of light) and remained skeptical of Einstein's theory as late as 1924. Even Millikan with an excellent background and training in the determination of both e (elementary electrical charge, Millikan, 1913b, see Chapter 7) and h (Planck's constant, Millikan, 1916b, see Chapter 8) could not fathom beyond the "observable empirical data" and give up his prior presuppositions. Ironically, at the age of 82, Millikan published his *Autobiography*, in which he reassured readers that his data in 1916 proved simply and irrefutably Einstein's theory. How can we explain this change in Millikan's thinking (even perhaps *volte-face*)? Such challenges provide a vivid image of science in the making.

Bending of light in the 1919 eclipse experiments: Einstein and Eddington. Evidence from the 1919 eclipse experiments (led by Eddington and Dyson) to provide support for Einstein's general theory of relativity has been the subject of considerable controversy (Chapter 9). A review of the literature shows that not only were the expeditions difficult to conduct, but the data produced also generated considerable amount of controversy even up to the present. A critical appraisal of the original data shows various contradictions. With this background let us go back to the dilemma faced by Dyson and Eddington and consider the following scenario: Eddington and Dyson are not aware of Einstein's General Theory of Relativity and particularly of the prediction that starlight near the sun would bend. Under these circumstances experimental evidence from all three sources (Sobral and Principe expeditions) would have been extremely uncertain, equivocal, and difficult to interpret. Eddington's presuppositions (he was fully aware of Einstein's General Theory of Relativity), however, led him to interpret the data as providing support for Einstein's theory. In the absence of presuppositions, none of the two theories (Newton's or Einstein's) could have been corroborated, and the most reasonable conclusion would have been that starlight is affected by the sun's gravitational field. It is plausible to suggest that Eddington's (Dyson et al., 1920) methodology approximated the traditional scientific method, which did not allow him to discard discrepant data. Dyson and especially Eddington were fully aware as to where the theory (Einstein's) was leading them. Nevertheless, given the inductivist/positivist milieu of their scientific community they could only be guided by their "implicit" presuppositions. Following any other alternative course involved the risk of having all their experimental data being declared ambiguous and ultimately invalid. This weighed more in this case as eclipse experiments are not easy to replicate (contrast this with Millikan's oil drop experiment, Chapter 7).

Lewis's covalent bond: From transfer of electrons to sharing of electrons. Gilbert N. Lewis's (1875–1955) theory of the covalent bond when first presented in 1916 was completely out of tune with established belief, according to which all bonds were formed by complete transfer of electrons from one atom to another (Chapter 10). Lewis (1916) is generally considered to have presented the first satisfactory model of the covalent (shared pair) bond based on the hypothetical/speculative cubic atom in 1916. The genesis of the cubic atom itself can be traced to an unpublished memorandum written by Lewis in 1902. The cubic atom led later to the formulation of the "rule of eight" or the "octet rule." From the standpoint of the polar theory (ionic bond) the idea that two negative electrons could attract each other or that two atoms could share electrons (covalent bond) was absurd, and seemed shocking to many chemists. No wonder, many chemists and textbooks ignore the historical antecedents and still consider the development of the covalent bond as a "Baconian inductive ascent" (cf. Chapter 1). According to this scheme, the following sequence facilitated the formulation of the covalent bond: finding that diatomic molecules and organic compounds cannot be understood by the ionic bond – postulation of the covalent bond by Lewis as an inductive generalization – theoretical explanation provided by Pauli's exclusion principle, as to how two electrons can occupy the same space. The crux of the issue is that this inductivist interpretation construes Pauli's exclusion principle as the theoretical explanation and ignores the fact that Lewis's hypothetical cubic atom was crucial for his later explanation of the sharing of electrons. A historical reconstruction shows how new ideas are resisted and that sharing of electrons (covalent bond) had to compete with a rival, the transfer of electrons (ionic bond).

Quantum mechanics: From Bohr to Bohm. The role of alternative interpretations in the history of science and in the development of quantum mechanics is important (Chapter 11). In order to facilitate understanding, Cushing (1991) has suggested that in the physical sciences scientific theories function at the following three levels: (a) *empirical adequacy* – consists of essentially in "getting the numbers right," in the sense of having a formula or an algorithm that is capable of reproducing observed data; (b) *explanation* – this is provided by a successful formalism with a set of equations and rules for its application; and (c) *understanding* – this is possible once we have an interpretation of the formalism that allows us to comprehend and to know the character of the phenomena and of the explanation offered. In the context of Cushing's scheme, it is plausible to suggest that Planck's contribution in 1900 looks more like at the level of empirical adequacy, whereas Einstein's contribution provided some degree of explanation, and Bohmian mechanics would facilitate understanding. Despite the importance of Bohm's (1917–1992) interpretation of "hidden variables" it was generally ignored in the 1960s, and is now being recognized as an important alternative. Among other reasons, it appears that Bohm's political affiliation, during the cold war, may also have played an important part. Bohm was fired from Princeton University because of his connections to the Communist party, which led him to exile, first in Brazil, then Israel, and finally England. These events had repercussions on his academic career and also with respect to the acceptance of his theory.

Wave-particle duality: De Broglie, Einstein, and Schrödinger. Postulation and understanding of wave–particle duality was a controversial topic from the very beginning and is closely enmeshed with the origin and development of the photoelectric effect and quantum theory (Chapter 12). J.J. Thomson (1925) in his *Structure of Light* compared the interplay between wave and particle theories of radiation to a struggle between a tiger and a shark in which each is supreme in his own element, but helpless in that of the other. Contributions of Einstein and de Broglie (1892–1987) in the postulation of wave–particle duality were considered to be crucial before any experimental evidence was available. Einstein supported de Broglie's ideas from the very beginning and considered that he had "lifted a corner of the great veil." It is interesting to note that de Broglie's theory of wave–particle duality not only preceded its experimental confirmation, but also suggested how experiments could be performed. These experiments were, however, difficult to perform and even leading physicists did not take up the challenge. Even when Davisson and Germer (1927a) performed experiments to support wave–particle duality, interpretation of the experimental data was extremely difficult. De Broglie's previous research, however, led him into controversies with two influential schools of physicists (Copenhagen and Munich), and this led Schrödinger (1887–1961) to develop the research program further. Despite Einstein's recommendation, Schrödinger was at first reluctant to accept de Broglie's ideas. Later, however, he acknowledged his debt to de Broglie. This shows how in scientific development, innovative and creative ideas are considered to be controversial and are resisted and sometimes even rejected due to the previous reputation of the scientist.

Searching for quarks: Perl's philosophy of speculative experiments. Martin Perl, a Nobel Laureate in physics, is at present working on the isolation of quarks (Chapter 13). Perl and his colleagues have used a Millikan-style methodology with improvements based on modern technology and outside the present experimental boundaries established by quantum chromodynamics. Given the difficulties involved in cutting-edge experimental work, he has designed a philosophy of speculative experiments, in which he outlines his research methodology that includes reason and speculations. Speculative experiments become important when the scientist is groping with difficulties, future of the research cannot be predicted and stakes are high due to competing groups (peer pressure). In his controversy with Ehrenhaft, Millikan's arguments were not precisely based on experimental evidence, but rather what he felt was the correct value of the elementary electrical charge, based on his presuppositions. Millikan considered many of his oil drops as beautiful and the possibility of fractional charges as too grotesque. Perl suggests that beauty is very desirable in speculative research and a beautiful experiment is always pleasant. Similarly, Polanyi (1964) suggested that our sense of scientific beauty responds to our vision of reality and suggests what data are intrinsically plausible and which empirical connections to reject as specious. Interestingly, all three (Millikan, Perl, and Polanyi) are experimental scientists and this may make our students wonder whether experimental data can be "grotesque" or "beautiful," and how that relates to our vision of reality.

Finally, a historical reconstruction of these episodes and experiments shows that interpretation of experimental data is difficult, which inevitably leads to alternative models/theories and thus facilitates the understanding of science as a human enterprise. At this stage it would be appropriate to pause and reflect as to why textbook authors, curriculum developers, and even scientists ignore the historical record and provide students a vision of science in the making that comes quite close to naivety.

Chapter 2
Quantitative Imperative Versus the Imperative of Presuppositions

Introduction

The quantitative imperative is the view that studying something scientifically means measuring it. W. Thomson (Baron Kelvin of Largs), a leading nineteenth-century British physicist, expressed this in succinct terms:

> When you can measure what you are speaking about and express it in numbers, you know something about it, and when you cannot measure it, when you cannot express it in numbers, your knowledge is of a meagre and unsatisfactory kind. It may be the beginning of knowledge, but you have scarcely in your thought advanced to the stage of a science. (Thomson, 1891, pp. 80–81)

Merton et al. (1984) have referred to this as the "Kelvin dictum," a methodological rule widespread amongst nineteenth-century scientists. Michell (2003) traces the origin of the "Kelvin dictum" in the history of science and concludes:

> Kelvin's dictum was not empty: it bespoke the considerable achievements of quantification.... Since antiquity quantification had always had a role, but following the work of Newton it not only occupied science's center-stage, it also dwarfed all other concerns. Kelvin's dictum expressed faith in the ultimate triumph of the quantitative vision of reality. (p. 8)

Interestingly, Kuhn (1977) in his essay, *The function of measurement in modern physical science*, starts by referring to the fact that the façade of the Social Science Research Building at the University of Chicago has the Kelvin dictum inscribed in the following terms: "If you cannot measure, your knowledge is meager and unsatisfactory" (p. 178). Such an inscription on a social science research building can lead to varied interpretations (cf. Merton et al., 1984). Kuhn (1977), however, used it to set the stage for his explanation of the role of quantitative and qualitative knowledge in the physical sciences. First, he asked a very pertinent question: Would such a statement (Kelvin dictum) be inscribed if it had been written, not by a physicist, but by a sociologist, political scientist, or economist? After responding in the negative, he attributes it to the prestige of modern physical science and the predominant role of measurement in such research, which is bolstered by a "profoundly unhistorical source, the science text." It is important to note that Kuhn even considers that textbooks are the sole source of most people's firsthand knowledge of the physical sciences. Next, he goes to considerable lengths to

M. Niaz, *Critical Appraisal of Physical Science as a Human Enterprise: Dynamics of Scientific Progress*, Science and Technology Education Library 36,

explain the mistaken belief that quantitative data lead to new inductive generalizations of laws and theories.

No wonder, Robert Andrews Millikan (1868–1953), perhaps one of the most successful experimental physicists of the twentieth century, also endorsed the Kelvin dictum. Actually, Millikan went beyond by subscribing to the pre-Socratic Pythagorean doctrine that the natural world is primarily quantitative in structure and concluded: "Indeed, from the point of view of that ancient philosopher [Pythagoras] the problem of all natural philosophy is to drive out qualitative conceptions and to replace them by quantitative relations" (Millikan, 1917, p. 4). This is rather ironic as Millikan's work on both the oil drop experiment and the photoelectric effect showed that "quantitative relations" do not displace "qualitative conceptions," but rather go hand in hand. (Oil drop experiment and the photoelectric effect are discussed in considerable detail in this monograph, see Chapters 7 and 8). Similarly, Stinner (1992) has argued cogently for understanding the relationship between experiment, hypothesis, and theory, within a framework based on history and philosophy of science.

According to Rigden and Stuewer (2005), in the physical sciences, the quantitative stands in sharp contrast to the qualitative. To understand any substantive topic, qualitative understanding is important, which requires a process of internalization so that an individual can draw on his resource of words to embrace a subject meaningfully. These authors illustrate this distinction between the qualitative and quantitative means of approaching a subject by referring to Einstein's general theory of relativity, completed in 1915 (see Chapter 9 for details). Based on his theory, Einstein predicted that light in the sun's gravitational field would be deviated through an angle of 1.75 arcseconds. In the 1919 eclipse expeditions, Arthur Eddington found that light did indeed deviate through an angle of 1.61 arcseconds. Eddington's interpretation of data was, however, controversial (cf. Chapter 9) precisely because he had a qualitative recourse (Einstein's theoretical framework) in order to understand the quantitative. This is precisely what Rigden and Stuewer (2005) have emphasized, namely that we need to have both the quantitative (data) and qualitative (theoretical ideas, in the case of Einstein also the image of free fall) means in order to understand science. No wonder these authors have asked a very pertinent question: Do physicists understand physics?

Despite the importance of the quantitative imperative it is important to recognize that progress in science, starting at least from Galileo (1564–1642), has gone through a constant process of conflict and controversy, in which the quantitative imperative was confronted by the imperative of presuppositions (Niaz, 2005a). Collingwood (1964, 1966) has referred to this as "enculturation" based on presuppositions that help us to interpret experimental data and understand scientific arguments as cogent or plausible. In other words, not all scientific theories and laws were strictly based on quantitative data (despite the rhetoric of the quantitative imperative), but rather there was a continual critical appraisal of data based on hypotheses and presuppositions. Based on the existing knowledge in a field of research a scientist formulates the guiding assumptions (Laudan et al., 1988), presuppositions (Holton, 1978, 1998), and "hard core" (Lakatos, 1970) of the research program that constitutes the imperative of presuppositions, which is not abandoned in the face of anomalous data.

At this stage it would be helpful to elaborate on Laudan et al.'s (1986) project of scrutinizing science. These authors follow the historicist approach to understanding science based on N.R. Hanson, T. Kuhn, I. Lakatos, P. Feyerabend, D. Campbell, S. Toulmin, W. Stegmueller, I.B. Cohen, G. Holton, D. Shapere, and L. Laudan, among others. Kuhn (1962) in his *Structure of Scientific Revolutions* has argued forcefully for abandoning our approach to history as a simple repository of anecdote and chronology if we really want to understand the dynamics of scientific progress. Laudan and his group consider the following paraphrase of Kant by Lakatos as an important guideline: philosophy of science without history of science is empty. Starting in the 1960s, the "historical school" has attempted to redraw and replace the positivist or logical empiricist image of science that dominated for the first half of the twentieth century. These studies have generally focused on some well-known case studies based on the chemical revolution, Bohr's atomic theory, and the Copernican revolution. Such scholarship has generally been well received in both the natural and the social sciences.

Laudan et al. (1986, 1988) have tried to go beyond by providing a more varied set of case studies, such as: Galileo, Newton, chemical revolution, Kekulé (benzene theory), fermentation theory, polywater episode, Ampère's electrodynamics, Brownian motion, Michelson–Morley experiment, plate tectonics, theory of an expanding earth, and Planck's quantum crisis, among others. This is an ambitious agenda and has provided a gold mine of information with respect to how theories are constructed, developed, and changed. The chemical revolution as presented in this volume by Perrin (1988) provides greater insight with respect to what was already known. Kuhn (1962) had previously used the chemical revolution to suggest that a change in guiding assumptions is a process that is both abrupt and total. Perrin's (1988) research, in contrast, reveals that Lavoisier's new guiding assumptions did not emerge full-blown but piece by piece over a decade or more.

Among other aspects, one that looms large in this whole study is that of "guiding assumptions" ("hard core" or "negative heuristic" for Lakatos; "paradigm" for Kuhn), and has considerable implications for the main thesis of this monograph. Different case studies provided an agreement with respect to the following claims about guiding assumptions (Donovan et al., 1988): (a) they are one of the most important units for understanding scientific change; (b) guiding assumptions, once accepted, are rarely abandoned just on the basis of negative empirical evidence (this is at odds with the Popperian insistence on the centrality of refutation, and even working scientists ignore that counterevidence does not provide instant rationality and thus change the theory); (c) observations are theory-laden; (d) guiding assumptions are not abandoned unless there is a new candidate to replace them; (e) coexistence of rival sets of guiding assumptions is the rule rather than the exception; (f) there is a constant debate and controversy among rival sets of assumptions; and (g) metaphysical and other nonscientific factors play an important role in the assessment of scientific theories and guiding assumptions.

It may be no surprise for someone following historians and philosophers of science that the thesis of Laudan et al. (1986, 1988) and Donovan et al. (1988) has been questioned and remains a subject of controversy. Howson (1990) considers this thesis to be nothing less than a "manifesto" of the "historical school" and still questions its methodology and some of the conclusions. The main argument is that looking for, and accepting, empirical evaluation through case studies amounts to a positivistic framework. Thus, negative empirical evidence would lead to rejection of the theses of the historical school. Without going into the merits of Howson's arguments, it is important to note that if science itself progresses through arguments, counterarguments, and controversies, no wonder philosophy of science has also chosen the same path. To make matters more difficult, Howson (1976) himself has also contributed to the development of the historical school.

More recently, Galison (2008) has expressed some skepticism with respect to historicism and its relationship to philosophy of science. He portrays three possible scenarios: (a) What if we cannot lean on the crutch of a preestablished philosophical framework into which the history of science is inserted? (b) What if we cannot take as a permanent given the positivist framework; and (c) What if we can no longer invoke Wittgenstein, Kuhn, or Pierce as the scaffolding on which historical detail is to be pitched? Galison recognizes that such a historicized project in which philosophy enters the stage with the history (and not before) would not only be difficult to write, but would also constitute "relentless historicism" (p. 123).

The objective of this monograph is to demonstrate that various experiments and episodes in the history of science involved the confrontation between the quantitative imperative and the imperative of presuppositions. In most cases the latter played a crucial role in the dynamics of scientific progress, which is illustrated in this chapter by the following aspects and episodes from the history of science:

1. Scientific laws as idealizations
2. Newton and the law of gravitation
3. Einstein and the Michelson–Morley experiment
4. Baconian "Inductive Ascent"
5. Role of heuristic principles in understanding nature of science (NOS)

Scientific Laws as Idealizations

According to Hanson (1958), "Why does motion cease? That was Galileo's problem" (p. 41). In contrast to Aristotle, who believed that a continually acting cause (i.e., force) was necessary to keep a body moving horizontally at a uniform velocity, Galileo predicted that if a perfectly round and smooth ball was rolled along a perfectly smooth horizontal endless plane there would be nothing to stop the ball (assuming no air resistance), and so it would roll on forever. Galileo, however, did not have the means to demonstrate that Aristotle was wrong, so he asked an epistemological question: What would make it (body) stop? Similarly, Galileo's discovery of the law of free fall later led to a general constructive model of falling bodies (Pascual-Leone, 1978). The law in its modern form can be represented by: $s = 1/2\ gt^2$ (s = distance, t = time, and g = a constant). In order to "prove" his law of free fall, Galileo should have presented empirical evidence to his contemporaries by demonstrating that bodies of different weight (but of the same material) fall at the same rate. If the leaning tower of Pisa mythical experiment (cf. Segre, 1989, for recent controversy) was ever conducted, it would have shown Galileo to be wrong. According to Pascual-Leone (1978), empirical computation of the value of *s* as a function of the variable *t*, "where vacuum and other *simplifying assumptions* are not satisfied" (emphasis added, p. 28), would lead to a rejection of the law. As a direct empirical test of Galileo's ideal law was not possible, he used his inclined plane experiment to show that as the angle of incidence approximated 90° (free fall), the acceleration of objects rolling down an inclined plane increasingly approximated a constant. According to Kitchener (1993, p. 142), by extrapolation one may assume it is also true of free fall as a limiting case (p. 142).

Following Galileo's method of idealization (considered to be at the heart of all modern physics by Cartwright, 1989, p. 188) scientific laws, being epistemological constructions, do not describe the behavior of actual bodies. According to Lewin (1935), for example, the law of falling bodies refers to only cases that are never realized, or only approximately realized. Only in experiment, that is under artificially constructed conditions (idealization), do cases occur that approximate the event with which the law is concerned. Furthermore, Lewin has argued that this conflict between quantification (Aristotelian) and qualitative understanding (Galilean) modes of thought constitute a paradox of empiricism. Galileo's law of free fall, Newton's laws, and gas laws all describe the behavior of ideal bodies that are abstractions from the evidence of experience and the laws are true only when a considerable number of disturbing factors, itemized in the *ceteris paribus* clauses, are eliminated (cf. Ellis, 1991; Matthews, 1987; McMullin, 1985; Niaz, 1999a). *Ceteris paribus* clauses play an important role in scientific progress, enabling us to solve complex problems by introducing simplifying assumptions (idealization). Lakatos (1970) has endorsed this position in the following terms: "Moreover, *one can easily argue that ceteris paribus clauses are not exceptions, but the rule in science*" (p. 102, original italics). Application of the Lakatosian methodology to Bohr's research program, as an example of how scientists progress from simple to complex models (simplifying assumptions), is quite instructive. Lakatos (1970) shows how Bohr used the methodology of idealization (i.e., simplifying assumptions) and developed the *positive heuristic* of Bohr's program by progressing from simple to complex models, that is, from a fixed proton-nucleus in a circular orbit, to elliptical orbits, to removal of restrictions (fixed nucleus and fixed plane), to inclusion of spin of the electron (this was not in discussion in 1913), and so on until the program could ultimately be extended to complicated atoms (Chapter 6 provides more details about Bohr's model and its historical reconstruction). This illustrates quite cogently the research methodology of idealization utilized for studying physical laws in particular and complex problems in general.

McMullin (1985) considers the manipulation of variables (disturbing factors) as an important characteristic of Galilean idealization:

> The move from the complexity of nature to the specially contrived order of the experiment is a form of idealization. The diversity of causes found in Nature is reduced and made manageable. The influence of impediments i.e., causal factors which affect the process under study in ways not at present of interest, is eliminated or lessened sufficiently that it may be ignored. (McMullin, 1985, p. 265)

Examples drawn from the history of science are presented to demonstrate how there was confrontation between the quantitative imperative and the latent imperative of presuppositions, which often led to controversy.

Newton and the Law of Gravitation

Newton's law of gravitation can be understood as "two bodies exert a force between each other which varies inversely as the square of the distance between them, and varies directly as the product of their masses" (Feynman, 1967, p. 14). This is how this law is presented to students in most parts of the world and then memorized and reproduced in examinations. According to Lakatos (1970) Newton's law of gravitation is one of the "best corroborated scientific theory of all times" (p. 92). Feynman (1967) endorses the view that it is "the greatest generalization achieved by the human mind" (p. 14). In spite of such impressive credentials, Cartwright (1983) asks "Does this law truly describe how bodies behave?" (p. 57) and responds laconically "Assuredly not" (p. 57). She explains further:

> For bodies which are both massive and charged, the law of universal gravitation and Coulomb's law (the law that gives the force between two charges) interact to determine the final force. But neither law by itself truly describes how the bodies behave. No charged objects will behave just as the law of universal gravitation says; and any massive objects will constitute a counterexample to Coulomb's law. *These two laws are not true: worse they are not even approximately true.* (Cartwright, 1983, p. 57, emphasis added)

Indeed, if scientific theories change and hence are tentative, this would perhaps be quite a surprise to many students and even teachers. This, however, is not the whole story. Feynman's version of the law of gravitation given above has an implicit *ceteris paribus* modifier or clause. According to Cartwright (1983), with the modifier the law would read: "*If there are no forces other than gravitational forces at work, then* two bodies exert a force between each other which varies inversely as the square of the distance between them, and varies directly as the product of their masses" (p. 58, modifier in italics). The law with the modifier, however, can explain in only very simple or ideal circumstances, and may even be considered as "irrelevant to the more complex and interesting situations" (Cartwright, 1983, p. 58).

In contrast to the reconstruction presented above, at the beginning of the twentieth century a leading historian of science conceptualized it in the following awe-inspiring terms:

> [T]he Newtonian gravitation formula has been generally accepted, and it still stands there as almost the only firmly established mathematical relation, expressive of a property of all matter, to which the *progress* of more than two centuries has added nothing, from which it has taken nothing away. (Merz, 1904, p. 384, emphasis added)

Indeed, only a year later Einstein had started to show how even a mathematical relation that had remained unchanged for 200 years also needed new formulations, such is the inexorable requirement of scientific progress and the tentative nature of science.

According to Lakatos (1970, p. 158), despite empirical success, ultimately all scientific theories turn out to be "false."

A recent appraisal of Newton's contribution is more sobering and provides a better understanding of his methodology:

> Our understanding of ... Newton's accomplishment, should therefore acknowledge the increasingly important role played by hypotheses. Thus Whewell urged that we should see Newton's declared rejection of hypothetical reasoning as no more than an expression of a tendency "prevalent in his time", and of his reaction to "the rash and illicit general assumptions of Descartes". (Gower, 1997, p. 131)

This leads to a dilemma: Did Newton formulate his law of gravitation based entirely on experimental observations (quantitative imperative, see Duhem in Chapter 3)? If the answer is in the affirmative, then he should have been aware that charged bodies would not follow the law of gravitation. Insight from Giere (1999) can help to resolve the dilemma:

> Most of the laws of mechanics as understood by Newton, for example, would have to be understood as containing the proviso that none of the bodies in question is carrying a net charge while moving in a magnetic field. That is not a proviso that Newton himself could possibly have formulated, but it would have to be understood as being regularly invoked by physicists working a century or more later. (Giere, 1999, p. 91)

Similarly, according to Kuhn (1977), when Newton enunciated his theory in the late seventeenth century, only his third law could be directly investigated by experiment. Convincing demonstration of the second law had to await the development of Atwood's machine, almost a century after the appearance of the *Principia*. Duhem (1914) suggested ways for testing Newton's first law of inertia. It specifies the behavior of those bodies which are under the influence of no impressed forces. However, no such body exists, as an observed body cannot be free of impressed forces (cf. Losee, 2001). Consequently, Newton's law of inertia cannot be a generalization about the observed motions of particular bodies. In other words, "Newtonian method" is more of an idealization or abstraction of such phenomena, and does not describe the behavior of actual bodies. Direct quantitative investigations of gravitational attraction proved even more difficult and were not available in the scientific literature until almost 1798. Finally, Kuhn has argued convincingly to conclude that it was precisely Newton's theoretical contributions that facilitated the construction of apparatus capable of facilitating experimental evidence. Positivist historians and textbooks of course convey a message that contradicts the historical record. Duhem (1914) was particularly critical of Newton and considered that the "Newtonian method," attractive as it may appear, was a dream (see Chapter 3 for details).

To recapitulate: it is plausible to suggest that if Newton did not entertain (or was not aware) that charge on a body could influence the force exerted between two bodies, then his quantitative data would have led to a different law of gravitation, viz., one that would include the following *ceteris paribus* (cf. previous section for details) clause "If there are no forces other than gravitational forces at work" and then the rest of the law as traditionally presented would follow. This clearly shows that in order to formulate his law of gravitation Newton inevitably resorted to idealization based on a hypothesis (despite claims to the contrary), that is a transaction between the quantitative imperative (data) and the imperative of presuppositions. This is precisely what Stinner (2000) has referred to as Newton's method, viz., the back and forth movement between mathematical models and experimental evidence. Similarly, Laudan et al. (1988, p. 7) have suggested that the soundness of theoretical claims and the significance of empirical data can be determined only by constantly moving back and forth between theory and experiment. This contrasts sharply with Newton's public stance of "hypotheses non fingo" (I do not feign hypotheses).

Einstein and the Michelson–Morley Experiment

Holton (1969a) has deplored the widely accepted myth (experimenticism) according to which progress in science is presented as the inexorable result of the pursuit of logically sound conclusions from experimentally indubitable premises (p. 980). The supposedly genetic relationship between the Michelson–Morley experiment (1887) and Einstein's special theory of relativity (STR) (1905) is an eloquent example of such a myth. Experimenticism, according to Holton (1969a), is the "unquestioned priority assigned to experiments and experimental data in the analysis of how scientists do their own work and how their work is incorporated into the public enterprise of science" (p. 977). The Michelson and Morley (1887) experiment, which provided a "null" result with respect to the ether-drift hypothesis, viz., no observable velocity of the earth with respect to the ether, is a good example of experimenticism. Many leading scientists and popular textbooks have generally attributed the origin of Einstein's special relativity theory in 1905 to the Michelson–Morley experiment. Details of the experiment and how it had to be repeated many times in order to provide consistent results go beyond the scope of this study (cf. Holton, 1969b; Lakatos, 1970; Laymon, 1988). Nevertheless, Einstein himself, despite some ambivalence, was quite forthright in setting the record straight and following is an example:

> In my own development Michelson's result has not had a considerable influence. I even do not remember if I knew of it at all when I wrote my first paper on the subject (1905). The explanation is that I was, for general reasons, firmly convinced that there does not exist absolute motion and my problem was only how this could be reconciled with our knowledge of electro-dynamics. One can therefore understand why in my personal struggle Michelson's experiment played no role or at least no decisive role. (Einstein's letter to Davenport, 9 February 1954, reproduced in Holton, 1969a, p. 969)

This shows clearly that Einstein was not particularly concerned as to how his presuppositions with respect to special relativity theory could be substantiated by experimental evidence, but rather what were the implications for electrodynamics. Nevertheless, some of the leading scientists themselves have contributed to perpetuate the myth with respect to a genetic relationship between the Michelson–Morley experiment and Einstein's special theory of relativity (STR). Millikan (1949) considered modern science to be essentially empirical (carefully tested experimental facts) and free of *a priori* conceptions and thus considered STR to be an inductive generalization based on the Michelson–Morley experiment. The experiment itself could hardly be considered as "crucial" or free of technical problems, as the debate with respect to the findings and their interpretation continued for almost 30 years (Holton, 1969a, b). Michelson himself considered the experiment to be a "failure" and subsequently used the interferometer (recognized as a sensitive scientific instrument) for measuring lengths. Michelson was awarded the Nobel Prize in 1907 not for providing evidence for STR, but rather for his optical precision instruments. Interestingly, Michelson himself did not mention the experiment in his *Nobel Lecture*. Actually, he was quite frustrated as the scientific community did not pay the attention that he thought the experiment deserved.

Another leading scientist, Joseph Petzoldt (a disciple of Ernst Mach), was even more emphatic in tracing the origin of Einstein's STR to the Michelson–Morley experiment. In the inaugural issue of *Zeitschrift für Positivistische Philosophie* (vol. 1, 1913), he wrote the lead article and minced no words in proclaiming the essence of positivism, "chief requirement of positivistic philosophy: greatest respect for the facts" (reproduced in Holton, 1969a, p. 977). In the same article, Petzoldt (1913) then argued cogently for the genetic link:

> There, one does not hesitate, for the sake of a single experiment, to undertake a complete reconstruction. The Michelson experiment is the cause and chief support of this reconstruction, namely the electrodynamic theory of relativity. To do justice to this experiment, one has no scruples to submit the foundation of theoretical physics as it has hitherto existed, namely Newtonian mechanics, to a profound transformation. (Reproduced in Holton, 1969b, p. 147)

This indeed provides an opportunity to scrutinize the positivist strategy of "respect for facts." A major problem with such a strategy is to determine "what are the facts." Interestingly, as early as 1906 (a year after Einstein published STR) W. Kaufmann, published in the same journal as Einstein, a categorical experimental disproof of STR (for details see Holton, 1969a, p. 976). Actually, Kaufmann went beyond and claimed that his experiments provided support not for Einstein but for rival theories of Abraham and Bucherer. What was Einstein's response? He questioned the fundamental assumptions of the rival theories and did not accept such "facts" (Kaufmann's data) as evidence against STR. To make matters worse, it was not until 1915 that Kaufmann's experimental equipment itself was found to be defective and hence the data obtained were questionable (for details see Holton, 1969a, b). It would be interesting to know if Petzoldt was aware of Kaufmann's original study published in 1906. History of science provides no answer to this. Nevertheless, this leads us to a dilemma: Can experimental data unambiguously lead us to the correct theory? This once again reflects the tension between the quantitative and qualitative imperative, quite similar to what another leading positivist scholar had referred to as "pristine" facts (Pearson, 1951) and Collingwood's (1964, 1966) presuppositions.

H. A. Lorentz (1895), a leading physicist, presented an auxiliary hypothesis to explain the null findings of the Michelson–Morley experiment. According to the ether theory the round-trip velocity of light should be lower in the direction of the earth's motion through the ether than in the direction perpendicular to this motion. In contrast, results of the Michelson–Morley experiment had shown that the round-trip velocity of light is the same in all directions on the earth's surface. The Lorentz contraction hypothesis suggested that all bodies on the earth undergo a minute contraction in the direction of the earth's motion through the surrounding ether. Popper (1959) considered that Lorentz had successfully explained the null results of the experiment.

Dayton Clarence Miller (a colleague of Michelson at Case) continued to work to provide experimental evidence for the ether-drift hypothesis. It was argued that the original Michelson–Morley experiment was conducted in the basement of a laboratory and hence the null result. Miller instead decided to perform the experiment on a hilltop (Mount Wilson observatory), between April 8 and 21, 1921, and found a nonzero drift (cf. Pais, 1982). On April 2, 1921, Einstein had arrived at Princeton to give lectures on the relativity theory. While still at Princeton, Einstein was informed of Miller's results and commented: "Subtle is the Lord, but malicious He is not" (reproduced in Pais, 1982, p. 113). Interestingly, the news correspondent of even a journal like *Science* reported to his readers that "Professor Miller's results knock out the relativity theory radically" (reproduced in Lakatos, 1970, p. 165). Holton (1969a) has considered such visions of science as "folkloric." For his part, Michelson (1927) continued to have doubts with respect to relativity, especially how the propagation of light waves without a medium can be explained (for details, see Shankland, 1963, 1964). Miller for his part continued to publish on the subject (Miller, 1933) and was one of the organizers of the international conference on the Michelson–Morley experiment held at the Mount Wilson Observatory (1927), in which among others Michelson and H.A. Lorentz participated.

So far we have studied the opinions of three leading physicists (Lorentz, Millikan, and Petzoldt) with respect to the Michelson–Morley experiment. Now let us consult a leading positivist philosopher of science, Reichenbach (1942, p. 51), who explicitly associated Einstein's theory and its origin to the Michelson–Morley experiment. Reichenbach was one of the most prominent philosophers who tried to convince the scientific community with respect to Einstein having based his theory on experimental facts. Gaston Bachelard (1949) enthusiastically claimed that relativity theory was born of an epistemological shock, due to the "failure" of the Michelson experiment, which roused classical mechanics from its dogmatic slumber. Again, Grünbaum (1963, p. 381) presents a similar thesis, in which he attributes to the experiment a crucial logical role in the genesis of relativity theory. Lakatos (1974) is critical of both Popper and Grünbaum for overestimating the negative result of the experiment. Such positivist presentations emphasize the importance of empirical data (Petzoldt's facts) with little concern for the circumstances that facilitated data (context of discovery), and recognizing only the logical analysis of the completed theory (context of justification). This tension between the logical and constructive aspects of a theory was explained by Einstein himself in eloquent terms at a ceremony to honor Michelson: "There is, of course, no logical way leading to the establishment of a theory, but only groping constructive attempts controlled by careful consideration of factual knowledge" (reproduced in Holton, 1969a, p. 979). Lakatos (1970, p. 162) has suggested that Einstein was unaware of Michelson, Fitzgerald, and Lorentz, but was stimulated primarily by Mach's criticism of Newtonian mechanics to postulate a new progressive research program. This program "predicted" and explained not only the outcome of the Michelson–Morley experiment, but also other hitherto unexplained facts. Interestingly, Laudan et al. (1986, p. 205), based on this episode have concluded that the correct prediction by a theory (Einstein's) of phenomena known (Michelson–Morley experiment) but unexplained prior to that theory counts strongly in favor of the theory.

After having presented the views of physicists and philosophers of science, now let us consult physics textbook authors. To our great surprise, Holton (1952) himself was also seduced by the genetic link between the Michelson–Morley experiment and Einstein's STR. Interestingly, Holton considered the experiment to be one of the "basic pillars" of the theory, and recognized his "error" by including an excerpt from his book along with other physics textbooks (for details, see Holton, 1969b, p. 141). Brush (2000) has analyzed 26 physics textbooks with respect to the relationship between the experiment and Einstein's STR. It is a relief to note that only nine textbooks still attributed Einstein's theory to the negative result of the Michelson–Morley experiment. Nevertheless, it is a cause for concern that these nine textbooks included some of the most well-known and widely used textbooks (not only in the USA, but almost all over the world), such as Serway (1996) and Sears et al. (1991). More recently, Arriassecq and Greca (2007) have reported that physics textbooks published in Argentina also suggest that the starting point for Einstein's STR was the Michelson–Morley experiment, which contributes to generate a distorted view of the dynamics of scientific research. This leads to the question: Why is the genetic relationship so seductive, not only for physicists, but also for philosophers of science and physics textbook authors? Addressing teachers, Holton (1969a) presented this dilemma in unequivocal terms: "Is our vision of the role of theory in physics today to some degree distorted by *an outdated or erroneous philosophy of science*?" (p. 969, emphasis added). To facilitate our understanding, Holton (1969a) has himself responded: Science textbooks place a high value on clear, unambiguous, inductive reasoning, which facilitates a clear genetic link from experiment to theory. No wonder, if eminent scientists (Millikan, Petzoldt) and philosophers of science (Reichenbach, Bachelard, Grünbaum) emphasize such links, textbook authors feel all the more reinforced in their inductive generalizations as the correct pedagogic strategy. Recent research has shown that such presentations are not an exception but rather quite frequent in physical science textbooks (Niaz, 2008a).

Cooper has emphasized that if we want students to understand why certain questions are asked at a particular time and why experiments are performed, then it is important to provide them with the *milieu of the time*, based on what the scientific community was thinking. Cooper has illustrated this with respect to the Michelson–Morley experiment:

> When a result is obtained can make all the difference. If, for example, the Michelson-Morley experiment had been done at the time of Copernicus, their (to them disappointing) result that they could measure no motion of the earth might have been greeted with a statement like "Why are you wasting your time? Everyone knows that the earth stands still at the center of the universe. Any attempt to measure its motion will give the answer you obtained: zero." Think of what effect this might have had on astronomers at the time of Copernicus. (Reproduced in Niaz et al., 2009)

This clearly shows the role of the scientific community and how its presuppositions are crucial in understanding an experimental finding.

It is not farfetched to suggest that the relationship between the experiment and STR was made popular after the theory had gained support among physicists, based on new developments. Foremost among these developments were the experimental findings of the bending of light in the 1919 eclipse experiments (Dyson et al., 1920) and the Compton effect (Stuewer, 1975). Support for Einstein's theory based on eclipse experiments was controversial and the experimental data themselves extremely difficult to interpret (this is discussed in detail in Chapter 9). According to Lakatos (1970, p. 162), starting from 1905, it took almost 25 years for the Michelson–Morley experiment to be understood and considered as the "greatest negative experiment in the history of science." Given the difficulties involved in interpreting experiments, Lakatos (1974), in general, is quite skeptical of the role of crucial experiments in theory construction.

Baconian "Inductive Ascent"

From about the middle of the nineteenth century spectroscopists had generated considerable amount of experimental data for atomic spectra, including hydrogen line spectrum. Johann Jacob Balmer (1825–1898), a Swiss school teacher, published an article in 1885 that was to have important implications for spectroscopic research and atomic theory (Balmer, 1885). Balmer's formula provided a means for obtaining the wavelength of any line in the hydrogen spectrum by multiplying a certain numerical factor (a fundamental number of hydrogen) by a series of fractions. Balmer, however, was not a mere numerologist, but had studied mathematics at various German and Swiss universities, before receiving his doctorate at Basel. It is interesting to note that Balmer is considered to be a devoted Pythagorean (Boorse & Motz, 1966) who believed that laws of the universe rest on the correlation of observed phenomena with the appropriate combination of numbers (cf. quantitative imperative, in this chapter).

Interpretation of Bohr's model of the atom within a history and philosophy of science perspective is instructive. Bohr's main objective was to explain the paradoxical stability of the Rutherford atom, and still most textbooks and even scientists consider Bohr's major contribution to be the explanation of the Balmer and Paschen series of the hydrogen line spectrum. According to Lakatos (1970), Bohr's problem was not to explain Balmer's and Paschen's series, but to explain the paradoxical stability of the Rutherford atom (p. 147). Moreover, Bohr had not even heard of these formulae before he wrote the "first version" of his paper. The "first version" is of course the "Rutherford Memorandum" written in June–July 1912 (cf. Heilbron & Kuhn, 1969, p. 244). A letter written by Bohr to Rutherford on January 31, 1913, shows that even then (first part of Bohr's trilogy was published in July 1913) he was not fully aware of the implications of spectroscopic research for his problem:

> I do not at all deal with the question of calculation of the frequencies corresponding to the lines in the visible spectrum. I have only tried, on the basis of the simple hypothesis, which I used from the beginning, to discuss the constitution of the atoms and molecules in their "permanent state." (Reproduced in Rosenfeld, 1963, pp. xxxvi–xxxvii)

Actually, Bohr was quite skeptical about the relevance of spectra for his model of the atom. Many years later, in an interview with Thomas Kuhn in 1962, Bohr expressed this quite explicitly: "The spectra was a very difficult problem. ... Just as if you have the wing of a butterfly, then certainly it is very regular with the colors and so on, but nobody thought that one could get the basis of biology from the coloring of the wing of a butterfly" (reproduced in Heilbron & Kuhn, 1969, p. 257).

Apparently, it was the spectroscopist H. M. Hansen who familiarized Bohr with the spectroscopic work and its implications for his model (cf. Jammer, 1996, p. 77). Having seen the importance of the spectra, Bohr is said to have repeated often: "As soon as I saw Balmer's formula, the whole thing was immediately clear to me" (reproduced in Rosenfeld, 1963, p. xxxix). Interestingly, Kuhn points out that even before the "Rutherford Memorandum" was discovered in Bohr's files, he had conjectured "that Bohr had developed a detailed, non-spectroscopic, quantized version of Rutherford's atom some time before he saw the relevance of the Balmer formula" (Heilbron & Kuhn, 1969, p. 255). Indeed, this clearly shows the importance of historical reconstructions for understanding how science progresses and is practiced.

Lakatos (1970, p. 147) has shown the importance of these events in the history of science in the following terms: Since the Balmer and the Paschen series were known before 1913 (year of Bohr's first publication), some historians present the story as an example of a Baconian "inductive ascent" consisting of three stages:

1. The chaos of spectrum lines, that is the accumulation of empirical data by spectroscopists before Balmer
2. An "empirical law" formulated by Balmer to obtain the wavelengths of lines in the hydrogen spectrum
3. The theoretical explanation provided by Bohr to explain electron transitions

From an inductivist perspective this constitutes a good example of the sequence: experimental data → empirical law → theoretical explanation, referred to as the "three floors of Whewell" or a Baconian "inductive ascent" by Lakatos (1970, p. 147). Despite all the work that has been done in the history and philosophy of science, most science curricula, textbooks, and even scientists still follow the inductivist perspective (Niaz, 2008a, b, c). Lakatos clearly uses this episode in the history of science to emphasize that science does not necessarily proceed from experimental observations to scientific laws and theories, through inductive generalizations. In spite of their many differences, most new philosophers of science would agree to this conceptualization of scientific progress (cf. Feyerabend, 1970; Hanson, 1958; Kuhn, 1970a, b; Lakatos, 1970; Laudan, 1977). This perspective based on the new philosophy of science can be summarized thus: "The role of observation is not to provide a firm basis from which generalizations can then be inductively extrapolated but, if anything, to provide some check on whether the promise of previously made theoretical commitments has been fulfilled" (Papineau, 1979, p. 100).

Once again, Lakatos (1970) expresses the argument cogently:

> [T]he progress of science would hardly have been delayed had we lacked the laudable trials and errors of the ingenious Swiss school-teacher [Balmer]: the speculative mainline of science, carried forward by bold speculations of Planck, Rutherford, Einstein and Bohr would have produced Balmer's results deductively, as test-statements of their theories, without Balmer's so-called "pioneering." In the rational reconstruction of science there is little reward for the pains of the discoverers of "naive conjectures." (p. 147)

At this stage it is instructive to study how philosophers of science have interpreted Bohr's model of the atom. According to Lakatos (1970):

> Bohr's problem was not to explain Balmer and Paschen series, but to explain the paradoxical stability of the Rutherford atom. Moreover, Bohr had not even heard of these formulae before he wrote the first version of his paper. (p. 147)

This version of the events has been corroborated by an extremely careful and detailed study by Heilbron and Kuhn (1969). Interestingly, most textbooks consider Bohr's major contribution to be the explanation of the Balmer and Paschen series of the hydrogen line spectrum. On the contrary, philosophers of science consider Bohr's explanation of the paradoxical stability of the Rutherford model of the atom as his major contribution (Heilbron & Kuhn, 1969; Lakatos, 1970). In a study based on 23 general chemistry textbooks, Niaz (1998) found that very few textbooks considered the explanation of the paradoxical stability of the Rutherford model of the atom as important and none of the textbooks interpreted the quantization of the Rutherford model within a historical perspective.

Lakatos (1970) goes on to show the importance of this event in the history of science:

> Since the Balmer and the Paschen series were known before 1913 [year of Bohr's first publication], some historians present the story as an example of a Baconian "inductive ascent": (1) the chaos of spectrum lines, (2) an empirical law (Balmer), (3) the theoretical explanation (Bohr). (p. 147)

A major premise of historians who follow the Baconian inductive ascent is that scientific theories and laws are primarily driven by experimental observations. Furthermore, such empiricist interpretations consider scientific progress to be dichotomous, viz., experimental observations lead to scientific laws, which later facilitate the elaboration of explanatory theories. Brush (1978) has explained cogently that "as soon as you start to look at how chemical theories developed and how they were related to experiments, you discover that the conventional wisdom about the empirical nature of chemistry is wrong. The history of chemistry cannot be used to indoctrinate students in Baconian methods" (p. 290). Similarly, Lakatos (1970) has argued that "the clash is not 'between theories and facts' but between two high-level theories: between an interpretative theory to provide the facts and an explanatory theory to explain them; and the interpretative theory may be on quite as high a level as the explanatory theory" (p. 129). In other words, scientific progress is characterized by a series of theories or models (plausible explanations) which vary in the degree to which they explain/interpret/predict the experimental findings.

Role of Heuristic Principles in Understanding the Nature of Science

Schwab (1962, 1974) made an important epistemological distinction between the methodological (empirical data) and interpretative (heuristic principles) components of scientific knowledge. According to Schwab (1974): "In physics, similarly, we did not know from the beginning that the properties of particles of matter are fundamental and determine the behavior of these particles; their relations to one another. It was not verified knowledge but a heuristic principle, needed to structure inquiry, that led us to investigate mass and charge and, later, spin" (p. 165). In other words, scientific progress does not depend on empirical data alone, but rather it is the heuristic principle (construction of the mind) that helps the scientist to look for relevant data (cf. Matthews, 1994; Monk & Osborne, 1997; Niaz, 1999a, b). Schwab's idea of a heuristic principle comes quite close to what modern philosophers of science have referred to as presuppositions (Holton, 1978)/hard core (Lakatos, 1970)/guiding assumptions (Laudan et al., 1988).

Given the complexity and multifaceted nature of the issues involved and a running controversy among philosophers of science themselves, implementation of nature of science (NOS) in the classroom has also been difficult. Despite the controversy a certain degree of consensus has been achieved within the science education community and the nature of science can be characterized, among others, by the following aspects (Abd-El-Khalick, 2004; Lederman, 2004; McComas et al., 1998; Niaz, 2008b; Smith & Scharmann, 1999):

1. Scientific knowledge relies heavily, but not entirely, on observation, experimental evidence, rational arguments, and skepticism.
2. Observations are theory-laden.
3. Science is tentative/fallible.
4. There is no one way to do science and hence no universal, recipe-like, step-by-step scientific method can be followed.
5. Laws and theories serve different roles in science and hence theories do not become laws even with additional evidence.
6. Scientific progress is characterized by competition among rival theories.
7. Different scientists can interpret the same experimental data in more than one way.
8. Development of scientific theories at times is based on inconsistent foundations.
9. Scientists require accurate record keeping, peer review, and replicability.
10. Scientists are creative and often resort to imagination and speculation.
11. Scientific ideas are affected by their social and historical milieu.

A review of the literature shows that most teachers in many parts of the world lack an adequate understanding of some or all of the different NOS aspects outlined above (Akerson et al., 2006; Bell et al., 2001; Blanco & Niaz, 1997; Clough, 2006; Lederman, 1992; Mellado et al., 2006; Pomeroy, 1993). This should be no surprise to anyone who has analyzed science curricula and textbooks, which have a pronounced stance towards an entirely empirical and positivist epistemology. Similarly, considerable amount of work has also been done to teach the nature of science in the classroom (Abd-El-Khalick & Akerson, 2004; Bianchini & Colburn, 2000; Irwin, 2000; Khishfe & Lederman, 2006; Lin & Chen, 2002). Nevertheless, the relationship between teachers' conceptions of NOS and their classroom practice is more complex than generally appreciated. Abd-El-Khalick and Lederman (2000) have attributed this to various factors, such as pressure to cover content, classroom management and organizational principles, concern for student abilities and motivation, institutional constraints, teaching experience, and difficulties in understanding the philosophical underpinnings of the nature of science. Despite the difficulties, research in science education has continued to work on the development and implementation of courses both at the undergraduate and high school levels, in order to facilitate students' and teachers' understanding of nature of science (NOS) (Abd-El-Khalick, 2005; Abd-El-Khalick et al., 1998; Pocoví, 2007). A major difficulty in implementing NOS is the expectation that students will come to understand it, by "doing science" (Lederman, 2004, p. 315). This is like assuming that students would come to understand photosynthesis just by watching a plant grow. In order to facilitate understanding of NOS, teachers need to go beyond the traditional curriculum and emphasize the difficulties faced by the scientists, and how interpretation of data is always problematic, leading to controversies among contending groups of researchers. An example of such a study is provided next.

In a recent study, Niaz (2008b) designed and implemented a course for graduate students (in-service teachers) based on historical controversies, in order to facilitate their understanding of nature of science. At the beginning of the course teachers were simply aware that ideas like the scientific method, objectivity, and empirical nature of science were considered to be controversial by philosophers of science. With this background, students were provided with reading material that dealt with various controversial episodes in the history of science and its implications for classroom instruction. Among others, some of the historical controversies discussed were the following: (a) Thomson's experiments were conducted to clarify the controversy with respect to the nature of cathode rays, that is, charged particle or waves in the ether (Based on Niaz, 1998; Niaz et al., 2002); (b) explanation of the large angle deflections of alpha particles was controversial: Thomson put forward the hypothesis of compound scattering (multitudes of small scatterings), whereas Rutherford propounded the hypothesis of single scattering based on a single encounter (based on Niaz, 1998; Niaz et al., 2002); (c) paradoxical stability of the Rutherford model of the atom, which led to Bohr's controversial thesis of incorporating Planck's "quantum of action" to the classical electrodynamics of Maxwell (based on Niaz, 1998; Niaz et al., 2002); and (d) Millikan–Ehrenhaft controversy with respect to the determination of the elementary electrical charge (based on Niaz, 2000b). These controversies refer to the atomic models of Thomson, Rutherford, and Bohr, and the determination of the elementary electrical charge (Millikan). All these topics formed part of the chemistry curriculum at the secondary and university freshmen level, and in-service teachers participating in the study were familiar with the topics. However, to their considerable surprise, although they were familiar with the topics, they were not aware of the difficulties involved in the elaboration of these models and the ensuing controversies. Presentation of these controversies led to lively debates during classroom discussions, both in favor of, and against, the different interpretations. Results obtained from the study facilitated in-service teachers' understanding of the following aspects of nature of science: (a) problematic nature of the scientific method, objectivity, and the empirical basis of science; (b) the role of speculation and controversy in the construction of knowledge based on episodes from the chemistry curriculum; (c) how did Bohr confirm his postulates? (This represents a novel way of conceptualizing Bohr's model of the atom based on a historical reconstruction (Lakatos, 1970), which goes beyond the treatment presented in most textbooks); (d) science does not develop by appealing to objectivity in an absolute sense, as creativity and presuppositions also play a crucial role; (e) differentiation between the idealized scientific law and the observations is crucial for understanding the complexity of science. One teacher expressed his/her evaluation of the course in the following terms: In the work of Thomson, Rutherford, Bohr, Millikan, and Ehrenhaft, besides logic, speculations played an important part. This reconstruction based on the history of science demonstrates that scientists adopt the methodology of idealization (simplifying assumptions) in order to solve complex problems. It is plausible to hypothesize that students adopt similar strategies in order to achieve conceptual understanding.

Abd-El-Khalick et al. (2008) have drawn attention to the importance of including nature of science (NOS) in high school chemistry textbooks. These authors analyzed 14 textbooks (published in the USA) including five "series" spanning 1–4 decades, with respect to the following NOS aspects: empirical, tentative, inferential, creative, theory-driven, myth of the scientific method, nature of scientific theories and laws, and social and cultural embeddedness of science. Results from this study revealed that chemistry textbooks fared poorly in their representation of NOS, which led the authors to conclude that "[t]hese trends are incommensurate with the discourse in national and international science education reform documents" (p. 1).

It is concluded that nature of science manifests in the different topics of the science curriculum as heuristic principles. Science education, by emphasizing not only the empirical nature of science but also the underlying heuristic principles, can facilitate conceptual understanding. Science curricula and textbooks, by emphasizing the historical context in which ideas, hypotheses, and theories develop, can be particularly helpful in facilitating conceptual understanding.

Finally, understanding of science inevitably leads to controversies with respect to interpretations based on different models/hypotheses/theories, viz., the imperative of presuppositions. Machamer et al. (2000) have on the one hand recognized the importance of controversies in scientific progress and on the other pointed out their paradoxical role:

> Many major steps in science, probably all dramatic changes, and most of the fundamental achievements of what we now take as the advancement or progress of scientific knowledge have been controversial and have involved some dispute or another. Scientific controversies are found throughout the history of science. This is so well known that it is trivial. What is not so obvious and deserves attention is a sort of paradoxical dissociation between science

> as actually practiced and science as perceived or depicted by both scientists and philosophers. While nobody would deny that science in the making has been replete with controversies, the same people often depict its essence or end product as free from disputes, as the uncontroversial rational human endeavor par excellence. (p. 3)

This discussion leads to another important issue: If observations can have varying interpretations, based on different theories and models which lead to controversies, can we conclude that this undermines the objective nature of science. To respond to this question we sought help from Leon Cooper (Nobel Laureate, physics, 1972), who responded in the following terms: "Observations can have varying interpretations, but this does not undermine the objective nature of science. It's somewhat ironic that what we like to call the meaning of a theory, its interpretation, is what changes. Think, for example, of the very different views of the world provided by quantum theory, general relativity and Newtonian theory" (reproduced in Niaz et al., 2009).

Chapter 3
Understanding Scientific Progress: From Duhem to Lakatos

Introduction

The influence among others of Hegel, Popper, and Kuhn on Lakatos has been recognized in the philosophy of science and mathematics literature. According to Urbach (1989): "Lakatos's methodology of science developed out of Kuhn's theory of paradigm, and this in turn was proposed largely in reaction to Popper's philosophy of science" (p. 399). Lakatos in his Ph.D. thesis recognizes his debt to three intellectual sources: Pólya's mathematical heuristics, Hegel's dialectic, and Popper's fallibilism (cf. Motterlini, 2002, p. 490). Similarly, the influence of Hegelian dialectics in Lakatos's philosophy of mathematics has been recognized by Ernest (1994) and Kadvany (2001). Ernest (1994) goes beyond and suggests a tension in the Hegelian and Popperian underpinnings in Lakatos's philosophy: "Under the sway of Popper and his followers at LSE in the 1960s Lakatos increasingly repudiated the Hegelian origins of his work. When former colleagues post-humously edited his works they removed further traces of Hegelian dialectics and also undermined Lakatos's fallibilism (most notably in Lakatos, 1976)" (p. 46, n. 13). Motterlini (2002) has suggested a similar tension between Hegelian historicism and Popperian fallibilism in Lakatos's philosophy. Thus, although, Lakatos's intellectual debt to Hegel, Kuhn, and Popper has been the subject of considerable research, the relation between Duhem and Lakatos has not been explored sufficiently. The objective of this chapter is to explore the various aspects of Lakatos's Methodology of Scientific Research Programs (MSRP) and suggest how Duhem could have provided a possible source of inspiration. This would facilitate a greater understanding of methodology in scientific progress. In his lectures on scientific method delivered at the London School of Economics in 1973, Lakatos (1999) endorsed Duhem's (1914) *The Aim and Structure of Physical Theory* in the following terms: "I recommend it to you as one of the best books written this century on the philosophy of science" (p. 46). Coming from Lakatos, a severe and overbearing critic of his contemporaries, this is quite a tribute. Gillies (2002), who worked with Lakatos for his doctoral thesis, has also recorded that he used to study Duhem with the very greatest attention. It is important to recognize that Lakatos (1970, p. 118) does not simply endorse

M. Niaz, *Critical Appraisal of Physical Science as a Human Enterprise: Dynamics of Scientific Progress*, Science and Technology Education Library 36,

or borrow from Duhem, but also provides a progressive problemshift in our understanding of philosophy of science. The objective of this chapter is to explore various aspects that are common to the philosophies of Duhem and Lakatos and suggest how the latter could have benefitted from the insights of Duhem.

A Brief Introduction to Pierre Duhem's Life and Career

Pierre Duhem (1861–1916) played an important part in French theoretical physics in the late nineteenth and early twentieth century. Later he made extensive studies of the history of physico-mathematical sciences and, as this chapter will show, is still considered to be an important philosopher of science (Crowe, 1990, 1999; De Broglie, 1953; Jaki, 1984; Martin, 1991). Born in Paris in 1861 he enrolled in Collège Stanislas, where he studied the physical sciences. At the age of 20, Duhem entered the prestigious École Normale Supérieure, Paris, and soon became interested in the study of the newly developing area of physical chemistry and thermodynamics, an interest that he maintained to the last. At the age of 23 he introduced the notion of thermodynamic potential and soon afterwards published a book, *Le potentiel thermodynamique et ses applications à la mécanique chimique et à la théorie des phénomènes électriques* (Paris, 1886). In 1887 he started his teaching career as a lecturer in the Faculty of Sciences of Lille University, where he taught thermodynamics, elasticity, and acoustics. During his 6 years in Lille he published almost 50 papers and 6 books including *Leçons sur l'électricité et magnétisme* (three volumes). In 1888, he completed his doctoral dissertation on the study of the theory of magnetism. This was his second doctoral thesis, the first one in physical chemistry having been turned down by the influential Marcellin Berthelot. At the age of 32 he became full professor in the Faculty of Sciences of Bordeaux University and kept this post until his death in 1916.

Duhem pursued his efforts at axiomatization and rigorous deduction of energetics as a form of abstract thermodynamics. He was strongly opposed to the idea of substituting for the formal arguments of energetics the uncertain models furnished by atomic theories and as such rejected the kinetic theory based on the work of Maxwell, Clausius, and Boltzmann. He admired Willard Gibbs and was one of the first to spread his work in France, for the rigor of his thermodynamic arguments, but did not agree with him on the atomic interpretation of thermodynamics based on general statistical mechanics. Similarly, his antipathy to atomic models prevented him from understanding the importance of Lorentz's theory of electrons.

Besides physics, Duhem devoted considerable amount of his time and effort to studying the history of science. In an important three-volume work, *Léonard de Vinci, ceux qu'il a lu et ceux qui l'ont lu*, he demonstrated that the great revival of mechanics, astronomy, and physics at the time of the Renaissance and in modern times had its roots in the intellectual work of the Middle Ages. This work presents discussions of a wide range of authors and that many of Leonardo's most important scientific ideas had their origin in medieval authors. Duhem's most important historical study was his ten-volume *Le système du monde: histoire des doctrines cosmologiques de Platon à Copernic*, whose first volume was published in 1913 and the last after his death in 1959. Thus, on the one hand he had good knowledge of physical theories and on the other an enormous erudition in the history of the sciences, which naturally led him to philosophy of science and his now classic *La théorie physique: son objet, sa structure*. His major contribution to philosophy of science is perhaps the Duhem–Quine thesis (Duhem, 1914; Quine, 1953, 1969). This thesis is generally presented as the radical underdetermination of scientific theories by experimental evidence and leads to the following two consequences: (a) there are no crucial experiments in science; and (b) it is not possible to localize defective hypotheses in scientific theories. Furthermore, the Duhem–Quine thesis continues to be the subject of considerable debate among philosophers of science (cf. Laudan, 1990; Weinert, 1995), and will be discussed later in this chapter.

Pierre Duhem is generally considered to be a positivist philosopher of science, agreeing with E. Mach that physical theory is above all an "economy of thought." Given his deep insight into understanding physical theories and even in agreement with some modern philosophers of science, it is rather paradoxical that he supported the school of "energetics" and rejected all attempts to introduce atomic theory. Being a staunch Catholic and politically a conservative, he attempted to separate physics from metaphysics, and give it a religious interpretation. Neither of these was a popular position in the secularist and liberal France of that period, and consequently he was frequently involved in controversy. Apparently, due to these two orientations he is considered as a positivist. Actually, he had a much more nuanced position with respect to physical theories as he constantly referred to the intricate relations between diverse appearances of reality, and not merely as a logical classification of observable phenomena. In 1894, he published a seminal paper in which he laid out his ideas on crucial experiments, and constitutes an important part of his famous *Aim and Structure of Physical Theory* (Duhem, 1906/1914).

Duhem, although an affable man, had an uncompromising character, which led him into controversies with his contemporaries and this explains why he never obtained a chair in an institution of higher education in Paris. Once he was offered the opportunity to teach history of science at the Collège de France, which he declined by reasoning that he was a physicist and not a historian. Despite these encounters with the intellectual elite of his days, 3 years before his death the Academy of Sciences in Paris appointed him as one of its nonresident members.

According to Crowe (1999), Duhem made a major contribution with respect to the following questions related to philosophy of science: (a) is inductivism (Duhem called it the "Newtonian method") the appropriate method for the exposition of physics? (b) are crucial experiments possible in physical science? and (c) is the law of inertia susceptible to experimental verification? Although these issues will be discussed in detail in this chapter (especially in relation to Lakatos), it is important to note that Duhem's answer to all three questions was in the negative. Of course, Duhem did not want to downplay the importance of experimental findings. His objective was to study the intricate relationship between the experimental findings and the scientists' theoretical framework or perhaps guiding assumptions (cf. Laudan et al., 1988, Chapter 2).

In the following sections various aspects that are common to the philosophies of Duhem and Lakatos are discussed, and possible relations are explored, especially with respect to:

1. Role of controversies
2. Role of presuppositions
3. Criterion of falsifiability: positive and negative heuristics of a research program
4. Role of contradictions (inconsistencies)
5. From progressive evolution to progressive problemshift
6. Duhem's dilemma: role of crucial experiments
7. Development of scientific theories
8. Duhem–Quine thesis within a Lakatosian framework
9. Conclusion

Role of Controversies

Duhem (1914, p. 217) has argued that pure logic is not always the best guide to make a choice between two theories. The reasons for selecting a theory, "which reason does not know," are based on "good sense":

> But these reasons of good sense do not impose themselves with the same implacable rigor that the prescriptions of logic do. There is something vague and uncertain about them; they do not reveal themselves at the same time with the same degree of clarity to all minds. Hence, the possibility of lengthy quarrels between the adherents of an old system and the partisans of a new doctrine, each camp claiming to have *good sense* on it side, each party finding the reasons of the adversary inadequate. The history of physics would furnish us with innumerable illustrations of these quarrels at all times and in all domains. (Duhem, 1914, pp. 217–218, italics added)

One such example cited by Duhem was how wave theory of light predicted that light moves faster in air than water, whereas the particle theory indicated the reverse. As an example, Duhem (1914, p. 218) cites the case of Biot who gave up supporting the emission hypotheses based on the particle theory (and not precisely on logical grounds, but rather "good sense"), after Foucault's experiment had shown that light traveled faster in air than in water.

Lakatos (1970) alludes to the Newton–Flamsteed (the first Astronomer Royal) controversy, in which "Newton constantly criticized and corrected Flamsteed's observational theories" (p. 131). Lakatos refers to an "appeal procedure" in case the theoretician (Newton) wishes to question the negative verdict of the experimentalist (Flamsteed). Based on this *appeal procedure*, Newton asked Flamsteed to reinterpret some of his data as they contradicted his theory. Based on such experiences, Lakatos (1970) concluded:

> *Even then, experience still remains, in an important sense, the 'impartial arbiter' of scientific controversy*. We cannot get rid of the problem of the 'empirical basis', if we want

> to learn from experience: but we can make our learning less dogmatic. (Original italics, p. 131)

Lakatos (1970) considered his *appeal procedure* to be a better alternative and critiqued Duhem in the following terms:

> Duhem adumbrated such standards [theoretical adjustments by which one is allowed to save a theory] in terms of 'simplicity' and 'good sense'. But *when* does lack of 'simplicity' in the protective belt of theoretical adjustments reach the point at which the theory *must* be abandoned? In what sense was Copernican theory, for instance, 'simpler' than Ptolemaic? The vague notion of Duhemian 'simplicity' leaves, as the naïve falsificationist correctly argued, the decision very much to taste and fashion. (Original italics, p. 117)

This discussion based on *good sense* (simplicity) and *appeal procedure* shows that both Duhem and Lakatos were grappling with the same issues, and Lakatos goes beyond by differentiating explicitly the *negative heuristic* and *protective belt* of a research program. In other words, the protective belt can be modified in the light of empirical evidence, whereas the negative heuristic is protected from refutation. Nevertheless, it is important to note that scientific controversies continue to provide considerable difficulties (cf. Niaz, 2005b for an appraisal of the oil drop experiment, and discussed in detail in Chapter 7) and a recent study provided the following insight: "While nobody would deny that science in the making has been replete with controversies, the same people often depict its essence or end product as free from disputes, as the uncontroversial endeavor par excellence" (Machamer et al., 2000, p. 3).

Role of Presuppositions

In order to understand the relationship between physical theory and experiment, Duhem (1914, pp. 190–200) goes into considerable detail with respect to the work of Newton and A.M. Ampère (1775–1836). A brief description of his views with respect to Newton and Ampère is necessary in order to understand the role of presuppositions. In the "General Scholium," which crowns Newton's *Principia*, he argued forcefully for all hypotheses to be drawn from experiment by induction, considered to be the "Newtonian method" (also see Chapter 2). Based on Kepler's laws, and as suggested by Newton, one can calculate the magnitude and direction of the forces between the sun (reference point) and the planets. Similarly, one can calculate the forces between the planets (reference point) and their satellites. These forces, however, are not the same as predicted by Newton because of perturbation between the two sets of forces, and consequently all heavenly bodies do not follow the orbit assigned to it by Kepler's laws. In other words, Newtonians will have to show that the observed perturbations are in agreement with the predictions. Based on these arguments, Duhem argued that Newton's law of universal gravitation cannot be derived inductively from the observational laws of Kepler, and concluded that if Newton's theory is correct, Kepler's laws are necessarily false. Duhem concluded that if the certainty of Newton's theory does not emanate from the certainty of Kepler's laws, how can the validity of this theory be established? (see Chapter 2 for more details on this point).

Next, Duhem points out that no other work in physical theory has been modeled more closely on the "Newtonian method" than André-Marie Ampère's (1775–1836) theory of electrodynamics. Ampère explicitly pointed out that in establishing these laws of electrodynamics, he had consulted only experience, and even included this aspect in the title of his classical *oeuvre, Théorie mathématique des phénomènes électrodynamiques uniquement déduites de l'expérience.* A major problem with Ampère's experiments was that he did not provide information with respect to how often were they repeated, how the conditions were modified, and the effect of these modifications. Such details were necessary to establish the reliability of the experiments and their absence led Duhem to conclude that the experiments were conducted with the "grossest degree of approximation." However, this is not the whole story. At the end of his treatise, Ampère (perhaps due to excess of candor) admitted that two of the instruments mentioned had not been constructed and hence the intended experiments had not yet been performed. According to Duhem, these two sets of apparatus (and the experiments based on them) constituted the main edifice of his findings, namely to determine the power of the distance according to which electrodynamic actions proceed. Apparently, besides his experiments, intuition also played an important role. At this stage, it is worthwhile to recall Millikan's methodology with respect to the oil drop experiment (1909–1916) in the determination of the elementary electrical charge, discussed in detail in Chapter 7.

In an effort to understand the relationship among experimental evidence, laws, hypotheses, and theories, Duhem (1914) asked a very pertinent question: "Does logic require our hypotheses to be simply experimental laws generalized by induction?" (p. 219). Duhem responded in very forthright terms:

> Now, we have recognized that it is impossible to construct a theory by a purely inductive method. Newton and Ampère failed in this, and yet these two mathematicians had boasted of allowing nothing in their systems which was not drawn entirely from experiment. Therefore, we shall not be averse to admitting among the fundamental bases of our physics postulates not furnished by experiment. (Duhem, 1914, p. 219)

Previously, in his book, Duhem (1914) had critiqued the Newtonian (inductive) method and provided details of how both Newton and Ampère espoused inductive generalizations and still based their theories on presuppositions, and concluded: "Hence, where Newton had failed, Ampère in his turn just stumbled. That is because two inevitable rocky reefs make the purely inductive course impracticable for the physicist" (p. 199).

Almost 60 years later, Lakatos in his lectures in 1973 recounted the details of Duhem's (1914) critique of the Newtonian method and provided the following thought-provoking insight: "What is curious is that, although *The Aim and Structure of Physical Theory* has been published in twenty languages and many editions since 1905, somehow nobody has ever discussed these painful points" (Lakatos, 1999, p. 47). Lakatos was of course ignoring that others (e.g., Otto Neurath) had already alluded to this (see note 11 by Motterlini, in Lakatos, 1999, p. 47). The influence of Duhem on Neurath has been explored by Cartwright et al. (1996). However, in the present context what is important to note is that Duhem's reconstruction of

Newton and Ampère perhaps provided the backdrop to Lakatos's conceptualization of the inductive method:

> Inductivism claims that a proposition is scientific if provable from facts; what we shall now set out to do is to show that no proposition whatsoever can be proven from facts. And certainly not any scientific proposition. (Lakatos, 1999, p. 36)

Let us now go back to Duhem to see how he generalized the experiences of Newton and Ampère in his philosophical framework:

> [B]efore he can judge whether the consequences of his hypothesis attain their object, before he can recognize whether they yield a methodic classification of, and a picture resembling, the experimental laws, *he must constitute the entire system from his presuppositions*; and when he asks logic to guide him in this difficult task, to designate which hypotheses he should choose, which he should reject, he receives merely this prescription to avoid contradiction, a prescription that is exasperating in the extreme latitude it allows to his hesitations. (Duhem, 1914, p. 221, italics added)

It is not farfetched to suggest that Duhem's dilemma with respect to constitution of the *entire system from his presuppositions* and the guidance from logic and other sources crystallized in Lakatos's *negative and positive heuristics of a research program*, which is the subject of the next section.

Criterion of Falsifiability: Positive and Negative Heuristics of a Research Program

Duhem (1914) expressed his dilemma with respect to the falsifiability of hypotheses or theories in the following terms:

> [T]he physicist can never subject an isolated hypothesis to experimental test, but only a whole group of hypotheses; when the experiment is in disagreement with his predictions, what he learns is that at least one of the hypotheses constituting this group is unacceptable and ought to be modified, but the experiment does not designate which one should be changed. (p. 187)

Now, Duhem was correct in emphasizing that we cannot take each one of the hypotheses in isolation and establish each one's validity through experiments. In other words, physical science is a system that must be taken as a whole. However, if we accept the Duhem thesis, it would be impossible to falsify an isolated hypothesis (Gillies, 2002, p. 16). Gillies (1993, Chapter 10) has provided an alternative interpretation of the Duhem thesis. Furthermore, if experimental evidence contradicts a group of hypotheses, how can we decide which of the hypotheses needs to be changed? This shows the problematic nature of the falsifiability criterion, espoused among others by Popper as well, and critiqued by Lakatos (1970). Considering that this criterion forms an important part of the Popperian philosophy of science, Lakatos (1999) considered it to be a "step back from Duhem" (p. 89). So, how do we go beyond Duhem? Lakatos (1970) suggested a solution by postulating that a research program has two important parts, the positive and the negative heuristics:

> The negative heuristic specifies the 'hard core' of the programme which is 'irrefutable' by the methodological decision of its protagonists [presuppositions]; the positive heuristic

> consists of a partially articulated set of suggestions or hints on how to change [guidance from logic and other sources], develop the 'refutable variants' of the research-programme, how to modify, sophisticate, the 'refutable' protective belt. (p. 135)

Precisely, Lakatos argued that in the face of contradictory experimental evidence, a scientist would not abandon those hypotheses that constitute the "hard core" of his research program. However, he can still learn from experience by incorporating changes in the positive heuristic or protective belt of the research program. To support Lakatos, a good example from the history of science is provided by Bohr's (1913a) research program. The "hard core" of this research program consisted of five postulates for which Bohr did not have convincing experimental evidence (besides being based in part on speculations) and still the scientific community (despite later criticisms) supported the publication of that research. Interestingly, Lakatos (1970) has suggested that given the contradictions in Bohr's 1913 paper (and despite Rutherford's recommendation), Popper would never have recommended its publication. Another example is provided by the advance of the perihelion of Mercury's orbit (an experimental evidence), which was not explained by Newtonian theory, but was only explained by Einstein's theory of relativity after more than 200 years (see Chapter 9). Issues related to falsificationism, refutation, confirmation, and verification have been hotly debated by philosophers of science. According to Giere (1988): "Lakatos turned Popper's falsificationist methodology on its head. For Popper, confirmations count for little; refutations are what matter. For Lakatos, refutations count for little; confirmations are what matter" (p. 39). Furthermore, according to Lakatos (1970, p. 137), verifications (i.e., confirmations) do not verify a program but rather show its heuristic/explanatory power. Duhem (1914) was even more explicit by pointing out that "[e]xperimental verifications are not the base of a theory, but its crown" (p. 204). In other words, in the initial phases of theory construction, presuppositions are more important rather than experimental findings. In the long run, however, verifications are more important, as they are the "crown" of a theory. Actually, Duhem (1914, p. 206) even suggested that in the initial phases the scientist is even free to ignore some of the experimental data (cf. Millikan's handling of the oil drops, Chapter 7).

Role of Contradictions (Inconsistencies)

Duhem (1914) differentiates between the empirical facts (experimental observations) and the theoretical framework of a physical theory and suggests that the former can be "true" or "false." However, with respect to

> propositions introduced by a theory, they are neither *true* nor *false*; they are only *convenient* or *inconvenient*. If the physicist judges it convenient to construct two different chapters of physics by means of hypotheses contradicting each other, he is free to do so ... to oblige physical theory to preserve a rigorous logical unity in its development would be to impose on the physicist's mind an unjust and intolerable tyranny. (p. 334)

Lakatos (1970) emphasizes the role of contradictions (inconsistencies) in much clearer terms:

> Indeed, *some of the most important research programmes in the history of science were grafted on to older programmes with which they were blatantly inconsistent*. For instance, Copernican astronomy was 'grafted' on to Aristotelian physics, Bohr's programme on to Maxwell's. Such 'grafts' are irrational for the justificationist and for the naïve falsificationist, neither of whom can countenance growth on inconsistent foundations. (p. 142, original italics)

Margenau (1950) expressed the same idea in picturesque terms: "[I]t is understandable that, in the excitement over its success, men overlooked a malformation in the theory's architecture; for Bohr's atom sat like a baroque tower upon the Gothic base of classical electrodynamics" (p. 311). Lakatos (1970) also reproduced this quote from Margenau and provided his own insight: "But as a matter of fact, the 'malformation' was not 'overlooked': everybody was aware of it, only they ignored it – more or less – during the progressive phase of the programme" (p. 142). No wonder, even students and teachers go through a process similar to that of the scientists in which some "progressive" parts of their thinking is inconsistently "grafted" on to a base that represents positivist thinking (cf. Blanco & Niaz, 1998, p. 353). According to Lakatos (1971), "if the game of science had been played according to Popper's rule book, Bohr's 1913 paper would never have been published because it was inconsistently grafted on to Maxwell's theory" (p. 113). Again, Lakatos goes beyond Duhem by suggesting that contradictions in new developing programs are more acceptable in scientific practice, especially during the "progressive phase," and Bohr's example is quite illustrative.

Finally, it is important to note the novel aspect of the separation in Lakatos's methodology (in contrast to Duhem) of the theoretical framework (negative heuristic) from the experimental observations (positive heuristic). Duhem's stipulation (observations can be true or false, however, theories are convenient or inconvenient and hence can live with contradictions) lacks the insight provided by "heuristic power" in Lakatos's (1970, p. 123) methodology. It is in this precise sense that Lakatos provides a progressive problemshift with respect to the formulations of Duhem. Interestingly, Lakatos (1970, p. 123) attributes the conflation between "proven truth" and "heuristic power" to both W. Whewell and Duhem. According to Lakatos, heuristic power provides the capacity to explain based on verifications, which do not necessarily "prove" a theory.

From Progressive Evolution to Progressive Problemshift

Duhem takes particular care to emphasize how hypotheses and theories are not the product of sudden creation, but the result of progressive evolution. He takes Newton's law of gravitation to illustrate that

> [t]hose who have a deeper insight into the history of physical theories know that in order to find the germ of this doctrine of universal gravitation, we must look among the systems of Greek science; they know the slow metamorphoses of this germ in the course of its millenary evolution; they enumerate the contributions of each century to the work which will receive its viable form from Newton; they do not forget the doubts and gropings through which Newton himself passed before producing a finished system; and at no moment in the

> history of universal attraction do they perceive any phenomenon resembling a sudden creation. (Duhem, 1914, p. 222)

In order to leave no doubt with respect to the progressive evolution of scientific theories (in contrast to product of a sudden creation), Duhem resorted to a metaphor from Greek mythology and concluded:

> The ordinary layman judges the birth of physical theories as the child the appearance of the chick. He believes that this fairy whom he calls by the name of science has touched with his magic wand the forehead of a man of genius and that the theory immediately appeared alive and complete, like Pallas Athena emerging fully armed from the forehead of Zeus. (Duhem, 1914, p. 221)

Interestingly, Lakatos (1970, p. 133), just like Duhem, also appraised the Newtonian research program and classified Newton's three laws of dynamics and the law of gravitation as part of the negative heuristic, which is irrefutable by the methodological decision of its protagonists. However, Lakatos goes beyond (progressive problemshift) by resorting to Greek mythology once again to explain that the "hard core" of a research program develops slowly:

> The actual hard core of a programme does not actually emerge fully armed like Athene from the head of Zeus. It develops slowly, by a long, preliminary process of trial and error. (Lakatos, 1970, p. 133, n. 4)

It is important to note that Duhem uses the metaphor (Athene from the head of Zeus) to describe the formation of scientific theories in general, whereas Lakatos uses it to understand the formation of the negative heuristic (hard core), which precisely fulfills Lakatos's (1970) requirement for a progressive problemshift. In other words, classification of a research program in negative and positive heuristics helps us to understand that just as the hard core is difficult to refute it also develops slowly and not like Athene from the head of Zeus. This insight in the development of scientific theories can be considered a novel idea (fact) and hence constitutes a progressive problemshift (cf. Lakatos, 1970, p. 118).

Duhem's Dilemma: Role of Crucial Experiments

In order to convince the reader that a crucial experiment is impossible in physics, Duhem (1914, p. 189) designed a didactic strategy based on two hypotheses concerning the nature of light: (a) Newton, Laplace, and Biot considered light to be consisting of projectiles hurled with extreme speed – according to this hypothesis light travels more quickly in water than in air; (b) Huygens, Young, and Fresnel considered light to be consisting of vibrations whose waves are propagated within an ether – according to this hypothesis light travels more quickly in air than in water. Foucault's experiment provided support for the second hypothesis, viz., the wave hypothesis. At this stage Duhem raises an important methodological issue that may be disconcerting even to our present-day physics students and textbook authors:

> [T]he debate is over; light is not a body, but a vibratory wave motion propagated by the ether; the emission hypothesis has had its day; the wave hypothesis has been put beyond

> doubt, and the crucial experiment has made it a new article of the scientific credo… how mistaken we should be to attribute to Foucault's experiment so simple a meaning and so decisive an importance; for it is not between two hypotheses, the emission and wave hypotheses, that Foucault's experiment judges trenchantly; *it decides rather between two sets of theories each of which has to be taken as a whole, i.e., between two entire systems, Newton's optics and Huygens' optics*. (Duhem, 1914, p. 189, emphasis added)

This was Duhem's dilemma: Foucault's experiment could not decide in favor of any of the two hypotheses, as both in turn formed part of a "bigger picture" (entire systems), and Duhem's philosophy had no mechanism to explain this. Duhem's *two entire systems* can be replaced and understood as two research programs within the Lakatosian framework. In other words, refutation of a hypothesis based on experimental evidence does not imply the same for the entire research program (system for Duhem). Duhem, of course, did not distinguish between the different parts of a research program, but Lakatos did. Furthermore, Lakatos (1970) goes beyond by suggesting that objective reasons for rejecting a research program are provided not by negative experimental evidence, but rather by a rival research program. In a section entitled *A New Look at Crucial Experiments: The End of Instant Rationality*, Lakatos (1970) concluded:

> *Can there be any objective* (as opposed to socio-psychological) *reason to reject a programme, that is, to eliminate its hard core and its programme for constructing protective belts?* Our answer, in outline, is that such an objective reason is provided by a rival research programme which explains the previous success of its rival and supersedes it by a further display of *heuristic power*. (p. 155, original italics, in a footnote Lakatos explains that by heuristic power he means explanatory power)

In the same section Lakatos (1970, pp. 158–159) explains why crucial experiments are seen to be crucial only decades later, and provides the following examples: (a) Kepler's ellipses were generally admitted as crucial evidence for Newton against Descartes only about 100 years after Newton's claim; (b) the anomalous behavior of Mercury's perihelion was known for decades and was accepted as evidence for the refutation of Newton's theory only after Einstein presented his general theory of relativity in 1915; (c) Young's double-slit experiment of 1802 was considered crucial between the corpuscular and wave theories much later based on the additional work of Fresnel; (d) Brownian motion was known for nearly a century before it was considered as crucial evidence in favor of the atomists; (e) Michelson–Morley experiment of 1887 was considered a negative crucial experiment for Einstein's theory of relativity after almost 25 years (Lakatos goes into considerable detail with respect to this experiment and devotes almost six pages; also see Chapter 2). This historical sweep, ranging from Kepler (1571–1630) to Einstein (1879–1955), provides on the one hand an opportunity to understand Duhem's dilemma and on the other a progressive problemshift.

Philosopher-physicist Gerald Holton has also attributed the conceptualization of crucial experiments to the "experimenticist fallacy" in which the sequence experiment to theory is imposed and concluded:

> [T]his attitude that has caused Einstein's supposed use of the Michelson result to have become a fixed part of the folkloric consensus about the history of science, a story as widely known and believed as the story of the falling apple in Newton's garden and of the two weights dropped from the Leaning Tower in Galileo's Pisa – two other cases in which

> experimental fact is supposed to have provided the genetic occasion for synthetic theory. (Holton, 1969a, p. 975; also cf. Holton, 1969b)

Kitcher (2000) has provided further support for the transition from Duhem to Lakatos in the following terms:

> The most primitive type of rationalism proposes that scientific controversies are resolved by designing and carrying out crucial experiments ... the experiment is run, nature speaks, and at least one of the disputants retires defeated. Unfortunately, as has become commonplace since Pierre Duhem, nature typically does not offer clear indictments. More sophisticated forms of rationalism attempt to show how assignments of credit and blame can be localized, perhaps by considering sequences of theories within a research program. (Lakatos, 1970, p. 21)

Sequences of theories within a research program thus provide a better means to decipher the degree to which experimental evidence can enrich our experience (positive heuristic) and still allow the scientific community to evaluate the far-reaching consequences, while the negative heuristic (hard core) of the research program is protected from refutation.

Development of Scientific Theories

This section is based on a critical appraisal of Duhem's conceptualization of scientific progress as dichotomous, viz., experimental observations lead to scientific laws, which later facilitate the elaboration of explanatory theories. In order to explore possible relations, differences, and contradictions between Duhem and Lakatos, a hypothetical dialogue based on their original writings is used as a historical reconstruction.

Duhem

> There is no doubt that several of the geniuses to whom we owe modern physics have built their theories in the hope of giving an explanation of natural phenomena, and that some even have believed they had gotten hold of this explanation ... we have to prove that the search for explanation is indeed the Ariadne's thread which had led them through the confusion of experimental laws and has allowed them to draw the plan of this labyrinth ... the descriptive part [based on facts] has developed on its own by the proper and autonomous methods of theoretical physics; the explanatory part has come to this fully formed organism and attached itself to it like a parasite. (Duhem, 1914, pp. 31–32)

Lakatos

> The problem is *not* what to do when 'theories' clash with 'facts'. Such a 'clash' is only suggested by the '*monotheoretical deductive model*'. Whether a proposition is a '*fact*' or a '*theory*' in the context of a test-situation depends on our methodological decision. (Lakatos, 1970, p. 129, original italics)

Duhem

When we analyze a theory created by a physicist who proposes to explain sensible appearances, we generally do not take long to recognize that this theory is formed of two really distinct parts: one is the simple representative part which proposes to classify laws; the other is the explanatory part which proposes to take hold of the reality underlying the phenomena. (Duhem, 1914, p. 32)

Lakatos

In the pluralistic model the clash is not 'between theories and facts' but between two high-level theories: between an *interpretative theory* to provide the facts and an *explanatory theory* to explain them; and the interpretative theory may be on quite as high a level as the explanatory theory ... *the problem is which theory to consider as the interpretative one which provides the 'hard' facts and which the explanatory one which 'tentatively' explains them.* In a mono-theoretical model we regard the higher-level theory as an *explanatory theory to be judged by the 'facts'* delivered from outside (by the authoritative experimentalist): in the case of a clash we reject the explanation.[1]

Duhem

It is not to this explanatory part that theory owes its power and fertility; far from it. Everything good in the theory, by virtue of which it appears as a natural classification and confers on it the power to anticipate experience, is found in the representative part [interpretative theory according to Lakatos] ... *whatever is false in the theory and contradicted by the facts is found above all in the explanatory part.* (Duhem, 1914, p. 32, italics added)

Lakatos

Hypothetical response from Lakatos, reconstructed by the author of this study: Dear Duhem: You sound very much like a naïve falsificationist, when you juxtapose facts and theories and give priority to the former. In view of the Duhem–Quine thesis,[2] for which you are justly famous, this is quite surprising. Let me put things into a perspective by citing two of your own statements on this issue:

The role of the scientist is not limited to creating a clear and precise language in which to express concrete facts; rather, it is the case that the creation of this language presupposes the creation of a physical theory. (Duhem, 1914, p. 151)

---

[1]The decision to use some monotheoretical model is clearly vital for the naïve falsificationist to enable him to reject a theory on the *sole* ground of experimental evidence. (Lakatos, 1970, p. 129, original italics)

[2]According to McMullin (1997), "Duhem has the unique distinction of having a 'thesis' named after him both in philosophy of science and in history of science" (p. 607).

> [W]hat the physicist states as the result of an experiment is not the recital of observed facts, but the interpretation and the transposing of these facts into the ideal, abstract, symbolic world created by the theories he regards as established. (Duhem, 1914, p. 159)

This hypothetical reconstruction[3] based on the writings of Duhem and Lakatos provides an insight into the issues both had to grapple. An important and salient aspect of this reconstruction is that both Duhem and Lakatos agreed that there are two types of physical theories: one provides the facts and the other the explanations. In the case of a clash between the two, Duhem suggested that the explanatory part acts like a "parasite," whereas Lakatos emphasized that hard experimental facts can only be understood in the context of a theoretical framework (cf. controversy between Newton and Flamsteed referred to by Lakatos, 1970, p. 131). Other examples of how the theory decides what can be considered as data are: (a) Millikan–Ehrenhaft controversy with respect to the oil drop experiment in the determination of the elementary electrical charge (Holton, 1978; Niaz, 2005b; also Chapter 7); and (b) controversial evidence of bending of light in the 1919 eclipse experiments to support Einstein's general theory of relativity (Earman & Glymour, 1980; Hudson, 2003; also Chapter 9). Interestingly, however, Duhem by upholding the experimental facts contradicts some of his own philosophical views in which he had recognized the importance of the scientists' theoretical presuppositions in order to interpret the data. It is plausible to suggest that given the positivist milieu (cf. Holton, 1992) of the early twentieth century, Duhem did recognize the dilemma but was not prepared to take the big leap, viz., recognize that theoretical frameworks are paramount in our understanding of the hard facts. Lakatos did make the leap and this precisely constitutes a progressive problemshift. At this stage it is important to note that the conceptualization of progress in science as dichotomous (laws leading to explanatory theories) is even today fairly widespread among science students, teachers, and textbooks (cf. Blanco & Niaz, 1997; Niaz, 2008a, b, c).

Duhem–Quine Thesis Within a Lakatosian Framework

Lakatos (1970) considered that this thesis can be understood to have two interpretations: (a) *weak interpretation*, that is, "it only asserts the impossibility of a direct experimental hit on a narrowly specified target and the logical possibility of shaping science in indefinitely many different ways" (p. 184); (b) *strong interpretation*, that is, "excludes any *rational* selection rule among the alternatives" (p. 184, original italics). According to Lakatos, Duhem seems to have held only the weak interpretation, whereas Quine supported the strong interpretation. Lakatos himself seems to favor the weak interpretation and suggested that the strong interpretation would be opposed by both the naive (Popperian) and the sophisticated

[3] Motterlini (1999) provides a similar historical reconstruction based on an imaginary dialogue between Feyerabend and Lakatos.

falsificationist (Lakatosian). There has been some discussion in the literature as well with respect to the differences between Duhem and Quine (cf. Cartwright, 1983; Darling, 2002; Laudan, 1996; Needham, 1998; Papineau, 1979). Furthermore, Lakatos (1970) clarified that the sophisticated falsificationist allows any part of the body of science to be replaced, only on the condition that it is replaced in a "progressive" way, so that the replacement successfully anticipates novel facts (p. 187). Lakatos (1970) goes beyond by explaining the Duhemian thesis in the following terms:

> An experiment, for Duhem, can never *alone* condemn an isolated theory (such as the hard core of a research programme): for such 'condemnation' we *also* need 'common sense', 'sagacity', and, indeed, good metaphysical instinct which leads us towards (or *to*) 'a certain supremely eminent order'. (Lakatos, 1970, p. 185, n. 1, original italics, underline added, expressions within single quotes are reproduced by Lakatos from Duhem, 1914)

Lakatos presents a clearer picture of the Duhemian interpretation of the Duhem–Quine thesis (underdetermination of scientific theories by experimental evidence), by alluding to the fact that an experiment cannot refute an isolated part of the theory. Lakatos goes beyond and provides a progressive problemshift by explaining that the isolated part precisely constitutes the hard core of a research program. Kitcher (2000) has made the same point cogently: "Anti-rationalists often suppose that rationalists are committed to the simplest position, and that mere invocation of Duhem or W.V. O. Quine is checkmate" (p. 21). In other words, the Duhem–Quine thesis does not protect the positive heuristic of a research program from refutation.

Conclusion

Exploration of the relation between Duhem and Lakatos shows that Lakatos not only endorsed Duhem, but also provided a progressive problemshift in our understanding of various aspects of philosophy of science. Following aspects substantiate this thesis:

(a) Both Duhem and Lakatos recognized the role of controversies in scientific progress. Lakatos critiqued Duhem for having attributed these difficulties to lack of "good sense" and "simplicity." Lakatos, on the contrary, introduced the idea of an "appeal procedure," by which the protective belt of the research program could be open to adjustments (based on the controversies), while the negative heuristic was protected.

(b) Both were equally critical of formulation of scientific theories and laws based on inductive generalizations. Lakatos resolved the tension in the Duhemian dilemma of presuppositions and guidance from empirical evidence by differentiating between the negative and positive heuristics of a research program. Presuppositions facilitate the formulation of the negative heuristic, whereas the positive heuristic can receive guidance from empirical evidence.

(c) Duhem recognized that development of scientific theories involved contradictions and stipulated that empirical facts are "true" or "false," whereas scientific theories can live with contradictions, as they were only "convenient' or 'inconvenient." Lakatos, on the contrary, emphasizes the role of controversies by suggesting that empirical facts provide heuristic power (capacity to explain), which do not necessarily "prove" a theory, and hence theories can progress even with contradictions.

(d) Duhem used the metaphor from Greek mythology (theories do not emerge like Athene from the head of Zeus) to conceptualize the progressive evolution of scientific theories in general. Lakatos, however, used the same metaphor to understand the formation of the negative heuristic (hard core) of a research program, which provides greater insight.

(e) Both are critical of the role of crucial experiments in the refutation of theories. Lakatos, however, goes beyond by suggesting that objective reasons for rejecting a research program are provided not by a crucial experiment, but rather by a rival research program.

(f) Both agreed that there are two types of physical theories: one provides the facts and the other the explanations. In the case of a clash between the two, Duhem suggested that the explanatory part acts like a "parasite," whereas Lakatos emphasized that hard experimental facts can only be understood in the context of a theoretical framework.

(g) Lakatos provides a clear picture of the Duhem–Quine thesis (weak interpretation) by alluding to the fact that an experiment cannot refute an isolated part of a theory, especially if this part constitutes the hard core of the research program.

Finally, it is important to emphasize that in spite of their differences (which are understandable, given their sociopolitical milieus and separation of time frame of almost 60 years), both Duhem and Lakatos conceptualized progress in science not merely as the accumulation of experimental data, but rather as dependent on creative imagination, as will be apparent from the following:

> Physical theory confers on us a certain knowledge of the external world which is irreducible to merely empirical knowledge; this knowledge comes neither from experiment nor from the mathematical procedures employed by theory, so that the merely logical dissection of theory cannot discover the fissure through which this knowledge is introduced into the structure of physics. (Duhem, 1914, p. 334)
>
> The direction of science is determined primarily by human creative imagination and not by the universe of facts which surround us. Creative imagination is likely to find corroborating novel evidence even for the most 'absurd' programme, if the search has sufficient drive. This look-out for *new confirming evidence* is perfectly permissible. Scientists dream up phantasies and then pursue a highly selective hunt for new facts which fit these phantasies. (Lakatos, 1970, p. 187, original italics)

Interestingly, Lakatos's remarks reproduced above were included in a section entitled *Duhem–Quine Thesis, and Its Relation to Falsificationism*. Duhem's vision of science in the making is indeed remarkable, given the positivist milieu of his day. From Duhem (1906, if we take the first edition) to Lakatos (1970), little over 60 years were necessary to understand science (a progressive problemshift) as clearly an human endeavor based on imagination and creativity. As an indicator of how scientific methodology has progressed it is remarkable that even physicists now

(27 years after Lakatos) recognize in public that "[c]hoices in the design of speculative experiments [cutting-edge] usually cannot be made simply on the basis of reason. The experimenter usually has to base her or his decision partly on what feels right, partly on what technology they like, and partly on what aspects of the speculations [presuppositions] they like" (Perl & Lee, 1997, p. 699; Martin Perl was the recipient of the 1995 Nobel Prize for Physics; also see Chapter 13). Indeed both Duhem and Lakatos, each in his own way and circumstances, prepared the terrain for Perl and the new generation of scientists.

Chapter 4
Kinetic Theory: Maxwell's Presuppositions

Introduction

The kinetic theory of gases has been the subject of considerable debate and controversy in the history and philosophy of scientific literature (cf. Achinstein, 1987, 1991; Brush, 1976; Clark, 1976; De Regt, 1996; Dorling, 1970; Elkana, 1974; Gavroglu, 1990; Kuhn, 1970a; Lakatos, 1970; Mendoza, 1975; Nyhof, 1988). The starting point of Maxwell's work on the kinetic theory of gases was his reading of the paper by Clausius (1857) entitled: "On the nature of the motion which we call heat" (cf. Garber et al., 1986, p. xix). Similarly, Maxwell (1860) recognizes the work of early kinetic theorists, such as Bernouilli, Herapath, Joule, and Krönig. Besides the influence of the early kinetic theorists, it seems that the work of greatest influence on Maxwell's development of the kinetic theory of gases may well have been the Essays by Adolphe Quetelet on the *Theory of Probabilities* in the early nineteenth century. Nevertheless, it is important to note that "statistical" for Maxwell in 1860 had the connotation of an emergence of regularity out of the apparently chaotic behavior of large numbers of molecules, which had little to do with its later recognition that the macroscopic gas laws are only probabilistic (cf. Porter, 1981, 1994). The role of presuppositions in the development of the kinetic theory, based on the following aspects, is explored in this chapter:

1. Clausius' simplifying (basic) assumptions
2. Maxwell's simplifying assumptions (presuppositions)
3. A Lakatosian interpretation of Maxwell's research program
4. "Progressive problemshifts" in Maxwell's research program
5. Van der Waals' equation of state: a "progressive problemshift"
6. Kinetic theory and chemical thermodynamics as rival research programs
7. Educational implications of the kinetic molecular theory of gases

Clausius' Simplifying (Basic) Assumptions

Clausius' (1857) is considered to be the first full-fledged kinetic theory of gases and his following basic assumptions can be considered as a prelude to those of Maxwell's: (1) the space actually filled by the molecules of the gas must be infinitesimal

M. Niaz, *Critical Appraisal of Physical Science as a Human Enterprise: Dynamics of Scientific Progress*, Science and Technology Education Library 36,

in comparison to the whole space occupied by the gas; (2) the duration of the impact (i.e., change of direction) of the molecules must be infinitesimal compared with the time interval between the collisions; and (3) the influence of the molecular forces between the molecules must be infinitesimal.

Maxwell's Simplifying Assumptions (Presuppositions)

Maxwell in his 1860 paper sets down the following simplifying (basic) assumptions of his theory (and thus extends those of Clausius), which have been summarized by Achinstein (1987, p. 410):

1. Gases are composed of minute particles in rapid motion.
2. Particles are perfectly elastic spheres.
3. Particles act on each other only during impact.
4. Motion of the particles is subject to mechanical principles of Newtonian mechanics.
5. Velocity of the particles increases with the temperature of the gas.
6. Particles move with uniform velocity in straight lines striking against the sides of the container, producing pressure.
7. Derivation of the distribution law assumes that the *x*-, *y*-, and *z*-components of velocity are independent.

According to Achinstein (1987): "How did Maxwell arrive at them [assumptions]? They are highly speculative, involving as they do the postulation of unobserved particles exhibiting unobserved motion" (p. 410). Did Maxwell have an independent warrant (i.e., plausibility of the hypotheses) for his simplifying assumptions? It is plausible to suggest that Maxwell's assumptions are precisely the *ceteris paribus* clauses (see Chapter 2), which helped him to progress from simple to complex models of the gases. Taking our cue from Galilean idealizations, it is plausible to interpret Maxwell's basic assumptions in the following terms: "The move from the complexity of nature to the specially contrived order of the experiment is a form of idealization. The diversity of causes found in Nature is reduced and made manageable. The influence of impediments, i.e., causal factors which affect the process under study in ways not at present of interest, is eliminated or lessened sufficiently that it may be ignored" (McMullin, 1985, p. 265). This research methodology of idealization, that is, building of simple to complex models, is an important characteristic of modern non-Aristotelian science (for details, see Kitchener, 1993; Kitcher, 1993; Matthews, 1987; Niaz, 1993). Lakatos (1970) has endorsed this position in the following terms: "Moreover, *one can easily argue that ceteris paribus clauses are not exceptions, but the rule in science*" (p. 102, original italics).

Maxwell's research program is yet another example of a program progressing on inconsistent foundations (similar to Bohr, cf., Lakatos, 1970, p. 142). Among other assumptions, Maxwell's (1860) paper was based on "strict mechanical principles" derived from Newtonian mechanics and yet at least two of Maxwell's simplifying assumptions (referring to the movement of particles and the consequent generation of pressure) were in contradiction with Newton's hypothesis explaining the gas laws based on repulsive forces between particles. Newton provided one of the first explanations of Boyle's law in his *Principia* (1687) in the following terms: "If a gas is composed of particles that exert repulsive forces on their neighbors, the magnitude of force being inversely as the distance, then the pressure will be inversely as the volume" (Brush, 1976, p. 13). Writing at the turn of the century, Merz (1904) attributed the importance of the kinetic theory, in contradistinction to the Newtonian (attracting and repelling forces) view, to the experiments conducted by Joule and Thomson, around 1853. Apparently, due to Newton's vast authority, Maxwell even in his 1875 paper, "On the dynamical evidence of the molecular constitution of bodies," reiterated that Newtonian principles were applicable to unobservable parts of bodies (cf. Achinstein, 1987, p. 418). Brush (1976) has pointed out the contradiction explicitly: "Newton's laws of mechanics were ultimately the basis of the kinetic theory of gases, though this theory had to compete with the repulsive theory attributed to Newton" (p. 14).

A Lakatosian Interpretation of Maxwell's Research Program

According to Lakatos (1970), the *negative heuristic* represents the "hard core" of the research program, consisting of simplifying (basic) assumptions considered "irrefutable" by the decision of the community of scientists. The *positive heuristic* represents the construction of a "protective belt" consisting of a "partially articulated set of suggestions or hints on how to change, develop the 'refutable variants' of the programme" (Lakatos, 1970, p. 135). The scientist lists anomalies, but as long as his research program sustains its momentum, he may freely put them aside, and "*[i]t is primarily the positive heuristic of his programme, not the anomalies, which dictate the choice of his problems*" (Lakatos, 1970, p. 99, original italics). A research program is progressing if it frequently succeeds in converting anomalies into successes, that is, if it is explainable by the theory – referred to as "*progressive problemshifts*."

Based on Lakatos' (1970) philosophy of science, Clark (1976) considers the seven simplifying assumptions (mentioned above) as the hard core (negative heuristic) of Maxwell's research program and summarizes it in the following terms: "[T]he behaviour and nature of substances is the *aggregate* of an enormously large number of very small and constantly moving elementary individuals subject to the laws of mechanics" (p. 45, original italics). Similarly, Clark (1976, p. 45) considers the following methodological directives as the positive heuristic (cf. Lakatos, 1970) of Maxwell's research program:

1. Make specific assumptions as to the nature of the elementary particles and as to their available degrees of freedom, such that all interactions among them are subject to the laws of mechanics.
2. All interactions shall be treated according to the laws of mechanics, while distribution of the properties of the molecular motion among the molecules shall be treated according to the laws of statistics.
3. Try to weaken or, if possible, eliminate the simplifying assumptions, so as to simulate, as far as possible, conditions obtaining in a "real" gas.

4. Use the specific assumptions introduced to investigate the internal properties of gases (e.g., viscosity) while the macroscopic (hydrodynamic) and equilibrium properties should be derivable as limiting cases.

"Progressive Problemshifts" in Maxwell's Research Program

Maxwell's major contribution (which was the first theory to do so) was to make predictions beyond the hydrodynamical laws (Boyle, Charles, Gay-Lussac, and others), referring to transport properties of gases. In subsequent work, Maxwell (1965) and others reduced/modified their original simplifying assumptions as formulated in the positive/negative heuristic of the program, in order to obtain:

(a) A more rigorous deduction of the law of velocities in a steady state.
(b) A better approximation of the effect of molecular collisions upon the values of transport coefficients (cf. Clark, 1976, p. 54).
(c) Instead of considering the particles as "elastic spheres" he introduced the concept of "centers of force," considered to be a "progressive problemshift" by Clark (1976, p. 62).
(d) In 1875, although Maxwell still maintained that Newtonian principles were applicable to understand the behavior of gases, he recognized the contradictory nature of one of his basic assumptions by pointing out that "[t]he pressure of a gas cannot therefore be explained by assuming repulsive forces between the particles" (Maxwell, 1875, p. 422).
(e) Boltzmann (1868, 1871, 1872) was particularly responsible for eliminating simplifying assumptions in Maxwell's theory by deriving: (i) Maxwell's law for the distribution of velocities for polyatomic molecules in an external field; and (ii) an expression for the velocity distribution in a layer of gas which is not in a steady state, thereby eliminating the need for Maxwell's assumption that the streaming velocity can simply be added to the steady state distribution.

van der Waals' Equation of State: A "Progressive Problemshift"

Development of van der Waals' (1873) thesis (continuity of the gaseous and liquid state) was based on the kinetic theory of gases, Clausius' virial theorem, and the experiments of Joule and Thomson, which showed that the temperature of a gas lowers as it expands. Van der Waals reasoned that the intermolecular forces which account for the cohesion of the liquid phase represented a property of the molecular model, and hence their effect should still be appreciable in the gaseous phase. It is interesting to observe that if Maxwell's simplifying assumptions were speculative (cf. Achinstein, 1987), van der Waals in a sense followed the same methodology, viz., "without elaborate justification. Its appropriateness should be judged by van der Waals' results rather than by a priori arguments" (Gavroglu, 1990, p. 219). Using his equation of state, van der Waals precisely reproduced Andrews' (1869) experimental results (isotherms for carbon dioxide), which demonstrated the continuity of the transition from the gaseous to the liquid state. The continuity of the transition implied a qualitative identity between two states of matter (liquid and gaseous) that appear quite different, and this identity led van der Waals to formulate his law of corresponding states in 1880.

The importance of van der Waals' work can be understood better if we consider his contribution as an attempt to reduce/modify the simplifying assumptions of Maxwell's theory. Clark (1976) interprets van der Waals' contribution in the following terms: "*What is important is that the novel predictions were found to be in good agreement with experiment. Application of the heuristic of the kinetic programme had resulted in empirical growth,* ['progressive problemshift'] in this case the discovery of a *new general law*" (p. 60, original italics). In spite of this retrospective positive evaluation of van der Waals' work by historians and philosophers of science, Maxwell's appraisal was surprisingly negative (cf. Maxwell, 1874a, b).

Kinetic Theory and Chemical Thermodynamics as Rival Research Programs

It is important to note that although in retrospect, these days we recognize the kinetic theory as one of the greatest achievements of nineteenth-century science, it was subject to considerable criticism by leading scientists of the day. For example, Ostwald (1927) criticized the kinetic theory for its "superficial habit to cover up rather than promote actual scientific tasks by *arbitrary assumptions* about atomic positions, motions and vibrations" (p. 178, emphasis added). It is interesting to observe that the very *simplifying assumptions* that helped Maxwell and recognized as the "hard core" of the research program by modern philosophers of science were considered to be "arbitrary" by some influential nineteenth-century scientists (especially E. Mach and W. Ostwald). It is not difficult to appreciate that the "hard core" of the kinetic theory's research program had a certain background to it, and to consider it as "arbitrary" was indeed a superficial criticism.

Duhem (1962), another important critic, questioned the atomic models used by the kinetic theory and critically appraised it with respect to chemical thermodynamics: "Thermodynamics had reached maturity and constitutional vigour when the kinetic hypothesis came along to bring it assistance it did not ask for, with which it has nothing to do and to which it owed nothing" (p. 95). Given Duhem's erudition in the history of science this statement is, to say the least, quite surprising (see Chapter 3 for Duhem's philosophy of science).

The opposition of many of these critics has been attributed to their philosophical approach to science: "[T]he rise of the school of 'Energetics' [thermodynamics] championed by Mach and Ostwald, represents an early attempt of the positivist philosophy to limit the scope of science. This school held that to use modern terminology the atom was not an 'observable', and that physical theories should not, therefore, make use of the concept" (Jaynes, 1967, p. 80). Brush (1974) has reasoned in a similar vein: "Those scientists who did suggest that the kinetic theory be abandoned in the later 19th century did so not because of empirical difficulties but because of a more deep seated purely philosophical objection. For those who believed in a positivist methodology, any theory based on invisible and undetectable atoms was unacceptable" (p. 1169). Brush (1976) goes on to emphasize:

> The leaders of this reaction, in the physical sciences, were Ernest Mach, Wilhelm Ostwald, Pierre Duhem, and Georg Helm. Mach recognized that atomic hypotheses could be useful in science but insisted, even as late as 1912, that atoms must not be considered to have a real existence. Ostwald, Duhem, and Helm, on the other hand, wanted to replace atomic theories by 'Energetics' (a generalized thermodynamics); they denied that kinetic theories had any value at all, even as hypotheses. (p. 245)

Einstein's criticism of Mach's and Ostwald's philosophical views clarifies the issues even further: "The prejudices of these scientists against the atomic theory can be undoubtedly attributed to their positivistic philosophical views. This is an interesting example of how philosophical prejudices hinder a correct interpretation of facts even by scientists with bold thinking and subtle intuition" (Quoted by Suvorov, 1966, p. 578).

For students and textbooks, it is important to note that although present-day kinetic theory and chemical thermodynamics in some sense do complement each other, the two developed for various decades as rival research programs (Niaz, 2000a, 2008a). This rivalry at times became a bitter dispute between the protagonists of the two research programs. Elkana (1974) has summarized the intellectual milieu of scientific circles in the late nineteenth century in eloquent terms:

> An important group of scientists led by Wilhelm Ostwald and Georg Helm developed in the 1870's and 1880's a metatheory of science, which later turned into an almost religious cult with the following hard-core metaphysics: all hypotheses should be banned from science and all observable phenomena should be reduced to one fundamental principle, namely the principle of energy. Its programme was to develop all scientific fields deductively from this unitary principle. It combined in its core the phenomenology of Kirchhoff, Mach and other great philosophers-scientists who all tried to eliminate untestable speculations from science. (p. 265)

Educational Implications of the Kinetic Molecular Theory of Gases

Based on the previous sections it is plausible to suggest that in order to understand kinetic theory, textbooks need to refer to the following heuristic principles (criteria).

Maxwell's Simplifying (Basic) Assumptions

Niaz (2000a) has shown that of the 22 general chemistry textbooks analyzed, 17 simply mentioned that the postulates of the kinetic theory were "assumptions" and only three described Maxwell's simplifying assumptions satisfactorily, and following is an example:

> At this point we want to build a model (theory) to explain *why* a gas behaves as it does ... laws do not tell us *why* nature behaves the way it does. Scientists try to answer this question by constructing theories (building models). The models in chemistry are speculations about how individual atoms or molecules ... cause the behavior of macroscopic systems (collections of atoms and molecules in large enough numbers so that we can observe them). A model is considered successful if it explains known behavior and predicts correctly the results of future experiments. But a model can never be proved absolutely true. In fact, by its very nature *any model is an approximation* and is doomed to fail at least in part. Models range from the simple (to predict approximate behavior) to the extraordinarily complex (to account precisely for observed behavior). ... A relatively simple model that attempts to explain the behavior of an ideal gas is the *kinetic molecular theory*. This model is based on speculations about the behavior of the individual particles (atoms or molecules) in a gas. (Zumdahl, 1990, pp. 434–435)

This presentation emphasizes speculative models and that models develop (tentativeness) in order to explain the behavior of gases. It can be argued that such presentations are not based on an overt understanding of the history and philosophy of science. Nevertheless, most teachers would agree that it constitutes a logical, understandable, and helpful preamble before presenting the postulates of the kinetic theory. Rodríguez and Niaz (2004a) have reported that of the 30 general physics textbooks analyzed, only two presented satisfactorily the role played by simplifying assumptions and five made a simple mention.

Inconsistent Nature of Maxwell's Research Program

All the general chemistry textbooks ignored that Maxwell's program, although successful, was also based on an inconsistent foundation (Niaz, 2000a), and following is an example: "From our knowledge of the gas laws and *Newtonian mechanics*, we can try to understand the bulk properties of a gas, such as pressure and temperature, in terms of the behavior of individual molecules" (Chang, 1981, p. 113, emphasis added). Many textbooks explicitly invoke Newtonian mechanics along with Maxwell's presentation of the kinetic theory, without realizing an inherent contradiction. Similarly, Rodríguez and Niaz (2004a) have reported that none of the general physics textbooks made a satisfactory presentation. It is important to note that most textbooks not only ignore that scientists face difficulties and manifest contradictions, but rather present the development of the kinetic theory as a simple accumulation of a series of theories. No wonder, students are at a loss to understand why we need yet another theory.

van der Waals' Contribution: Reducing/Modifying Basic Assumptions

Niaz (2000a) has reported that 14 general chemistry textbooks described satisfactorily van der Waals' contribution as an attempt to reduce/modify the basic assumptions. Following is an example of a satisfactory description:

> Van der Waals examined critically the postulates of the kinetic molecular theory and recognized that some of these postulates had to be modified if the kinetic theory was to account more accurately for the behavior of real gases. (Quagliano & Vallarino, 1969, p. 140)

Most textbooks do not conceptualize van der Waals' contribution as an attempt to modify Maxwell's simplifying assumptions, which led to a "progressive problemshift" (Lakatos, 1970). Brush (1976) has highlighted this aspect in the following terms: "Thus, 100 years after its original publication, van der Waals' theory still serves as an illuminating example of how an astute scientist can penetrate to the heart of an important but complex phenomenon by the proper choice of *simplifying approximations*, thereby opening up a new field of theoretical and experimental research" (p. 251, emphasis added). Similarly, Rodríguez and Niaz (2004a) have reported that only one general physics textbook made a satisfactory presentation and five simply mentioned it.

Kinetic Theory and Chemical Thermodynamics as Rival Research Programs

Niaz (2000a) has reported that none of the general chemistry textbooks described this rivalry satisfactorily and only three made a simple mention. Following is an example of a textbook that made a brief mention:

> In addition to his extensive calculations in the kinetic theory of gases, he [Boltzmann] laid the foundations of the field of statistical mechanics, and explained the second law of thermodynamics on the basis of the atomic theory of matter. Boltzmann's work in statistical mechanics, now universally recognized, was strongly attacked by Wilhelm Ostwald (see Chapter 9) and others who did not believe in atoms. (Segal, 1989, p. 125)

Interestingly, in Chapter 9, this textbook mentions that Ostwald was probably the last great chemist who refused to believe in atoms, and tried to explain all material phenomena by considering energy changes (energetics). By ignoring the rivalry between kinetic theory and chemical thermodynamics, textbooks present the development of kinetic theory as a smooth process that involved no intellectual conflicts. On the contrary, according to Lakatos (1970) history of science has been characterized by conflicts between rival research programs.

Understanding Behavior of Gases: From Algorithmic to Conceptual

None of the textbooks described satisfactorily or briefly mentioned the two modes of solving gas problems, viz., the algorithmic mode and that of conceptual understanding (Niaz, 2000a). It seems that textbooks do not visualize the transition from hydrodynamical laws (e.g., Boyle) to kinetic theory (Maxwell and Boltzmann) as an opportunity to increase our conceptual understanding of gaseous behavior. Some textbooks did include problems that could be considered as conceptual; however, they lacked the framework based on the two modes of understanding gases. Similarly, Rodríguez and Niaz (2004a) found that none of the general physics textbooks explicitly recognized the difference between the two modes of solving gas problems.

It seems plausible to suggest that resolution of gas problems based on the Ideal Gas Law, derived by the inductive process, primarily requires manipulation (i.e., enumeration of particulars) of the different variables of the ideal gas equation ($PV = nRT$), and thus can be characterized by the "algorithmic mode." On the other hand, resolution of gas problems based on the Ideal Gas Law, which derives its meaning from the kinetic molecular theory of Maxwell and Boltzmann (a hypothetico-deductive system), requires the understanding of a pattern within which data appear intelligible, that is, a sort of "conceptual gestalt" (Hanson, 1958, p. 90). If the nature and methods of science influence the history of science, then it seems that this epistemological perspective has instructional significance (for details, see Niaz & Robinson, 1992). Interestingly, Hanson (1958) has pointed out that in the history of science the empirical work of those like Boyle, Cavendish, Ampère, Coulomb, and Faraday, although indispensable, at times is overestimated, in comparison to those who provide an explanation of the empirical data (e.g., Maxwell and Boltzmann), that is, "conceptual gestalt."

Finally, this chapter shows that most textbooks ignore an essential aspect of scientific progress, viz., how the diversity of elements found in the behavior of gases is studied/controlled through *ceteris paribus* clauses (approximations), leading to tentative theories that increase in their explanatory/heuristic power. Even when textbooks present historical details, it invariably is in the form of names of famous scientists including their pictures, year of work, and anecdotes. It appears that textbooks ignore historical details, not due to limitations of space, but rather due to a lack of a history and philosophy of science framework. Maxwell and Boltzmann facilitated our understanding of gas behavior beyond the observable hydrodynamical laws, by explaining and predicting the internal properties of gases. It is plausible to suggest that this transition provides a blueprint for similar changes in the understanding of students. The fact that most of our students can manipulate equations, find derivatives, and apply algorithms (cf. Rosnick & Clement, 1980), and yet fail to comprehend qualitative description of real-world everyday problems is an indicator of an educational system that focuses on manipulative skills.

Chapter 5
Periodic Table of the Chemical Elements: From Mendeleev to Moseley

Introduction

From a history and philosophy of science perspective the periodic table of elements has generally been considered as a classification, system, table, law, and very rarely as a theory. According to a historian of chemistry: "From the 1870s Mendeleev's Periodic Table came to adorn every chemical lecture room; it compressed a great deal of knowledge into a small compass, meaning that the student no longer had to be burdened with a great load of unrelated brute facts" (Knight, 1998, p. xii). Another historian goes further by recognizing that the periodic table "has contributed much more than mere classification. It has been a conceptual tool which has predicted new elements, predicted unrecognized relationships, served as a corrective device, and fulfilled a unique role as a memory and organization device" (Ihde, 1969, p. ix). Van Spronsen (1969) presents a detailed account of various attempts to classify elements between 1817 and 1860. However, a major problem with such classifications was that the atomic weights were not yet determined correctly and nor were they understood well, because "Dalton's atomic theory was too recent to have been conclusively demonstrated" (van Spronsen, 1969, p. 95).

Most historians consider the International Congress held in Karlsruhe (3–5 September 1860) as crucial in the development of chemistry and the periodic table in particular. A circular (dated July 10, 1860) sent by the organizers of the Congress to most outstanding chemists of Europe outlined its objective as the need to reach a consensus on "[m]ore precise definitions of the concepts of atom, molecule, equivalent, atomicity, alkalinity, etc.; discussion on the true equivalents of bodies and their formulas; initiation of a plan for a rational nomenclature" (reproduced in De Milt, 1951, p. 421). Mendeleev attended the Congress and was greatly impressed by Cannizaro's contribution and in a letter dated September 7, 1860, he summarized an important achievement of the Congress:

> It is decided to take a different understanding of molecules and atoms, considering as a molecule the amount of a substance entering a reaction and determining physical properties, and considering as an atom the smallest amount of a substance included in a molecule. Further, it reached an understanding about equivalents, considered as empirical, not depending on the understanding about atoms and molecules. (Reproduced in De Milt, 1951, p. 422)

M. Niaz, *Critical Appraisal of Physical Science as a Human Enterprise: Dynamics of Scientific Progress*, Science and Technology Education Library 36,

This clearly shows the tension that was to reflect for many years in Mendeleev's later writings, viz., on the one hand he wanted to understand the nature of atoms and molecules (i.e., atomic theory), and on the other he espoused the use of equivalent weights based on empirical findings. Gordin (2002) also recognizes the importance of the Karlsruhe Congress in the development of Mendeleev's thought on the periodic system. In my opinion, the development of the periodic table is closely related to the atomic theory, role of empirical evidence in the development of theories, and whether the periodic table itself was a simple classification/codification scheme, empirical law, or a theory. Given these antecedents and the long history of the periodic table and its relevance for chemistry and chemistry education, it is not surprising that historians and philosophers of science are far from reaching a consensus with respect to its origin, nature, and development (Bensaude-Vincent, 1986; Brush, 1996; Christie, 1994; Christie & Christie, 2003; Dmitriev, 2004; Gordin, 2004; Hettema & Kuipers, 1988; Howson & Franklin, 1991; Kaji, 2003; Lipton, 1991; Maher, 1988; Shapere, 1977; Vihalemm, 2003; Wartofsky, 1968; Weisberg, 2007; Ziman, 1978).

According to Brush (1996) scientists generally propose a hypothesis, deduce its consequences, make predictions, and do experiments to see if the predictions are borne out. Ziman considers that the "fundamental purpose of science is to acquire the means for reliable prediction" (1978, p. 32). Although actual scientific practice is much more complex and controversial, van Spronsen (1969) does recognize the importance of Mendeleev's prediction of gallium, after it was discovered in 1875. The objective of this chapter is to elucidate the nature of Mendeleev's contribution to the origin, nature, and development of the periodic table, and this reconstruction is presented in the following sections:

1. Periodicity in the periodic table as a function of atomic theory
2. Role of predictions in scientific theories and their implications for the periodic table
3. Relative importance of accommodations and predictions as possible support for Mendeleev's periodic table
4. Mendeleev's contribution: theory or an empirical law?
5. Contributions of Thomson, Lewis, Bohr, and Moseley
6. Mendeleev's Periodic Law: does it follow a "Baconian Inductive Ascent"?
7. Educational implications of a historical reconstruction of the periodic table

Periodicity in the Periodic Table as a Function of the Atomic Theory

Many students must have wondered how a simple arrangement of the chemical elements could provide predictive and explanatory (accommodation) power to Mendeleev's periodic table. Many chemistry textbooks (Brito et al., 2005) and teachers even suggest and give the impression that the problems associated with the periodic table could not be solved until Moseley (1913, 1914) and others showed that elements need to be ordered according to their atomic number. According to van Spronsen (1969):

> The actual development of the periodic system seemed to require a catalyst! We think it proper to attribute this catalytic action to Cannizaro's famous Karlsruhe lecture at the 1860 Congress. He *made the distinction between atoms and molecules and defined such concepts as valence*; in our opinion this initiated the second stage of the discovery and started the history proper of the periodic system of chemical elements. (p. 1, emphasis added)

In spite of this fairly categorical statement with respect to the role played by the atomic theory by a major historian of the periodic table, many historians attribute its success primarily to empirically observed properties of the elements (inductive generalization). It is important to recall that most of the pioneering work of Mendeleev was conducted from 1869 to 1889, before Thomson (1897), Rutherford (1911), Bohr (1913), and Moseley (1913, 1914) laid the foundations of the modern atomic theory. So how could Mendeleev conceptualize periodicity as a function of the atomic theory? An answer to this question will precisely show Mendeleev's ingenuity, farsightedness, creativity, and the ability to "speculate." Despite Mendeleev's own ambivalence and ambiguity, a historical reconstruction does provide a convincing story of this remarkable contribution to our knowledge. Before presenting the reconstruction it is important to note that Mendeleev had the following important sources of information: Dalton's atomic theory, law of multiple proportions, Cannizaro's Karlsruhe lecture, fairly reliable atomic weights, atomicity (valence), and various physical and chemical properties of the elements.

Step 1: Even in his first publication Mendeleev referred to the relationship, albeit implicitly, between periodicity, atomic weights, and valence: "The arrangement according to atomic weight corresponds to the *valence* of the element and to a certain extent the difference in chemical behavior, for example Li, Be, B, C, N, O, F" (Mendeleev, 1869, p. 405, original emphasis).

Step 2: After the discovery of gallium and scandium, Mendeleev expressed the relationship between atomic weight and atomic theory much more explicitly: "It is by studying them [atomic and molecular weights], more than by any other means, that we can *conceive the idea of an atom and of a molecule*. By this fact alone we are enabled to perceive the great influence that studies carried on in this direction can exercise on the progress of chemistry.... The expression atomic weight* implies, it is true, the hypothesis of the atomic structure of bodies" (Mendeleev, 1879, p. 243, emphasis added. The asterisk leads the reader to the following footnote: "By replacing the expression of atomic weight by that of elementary weight, I think we should, in the case of elements, avoid the conception of atoms"). This footnote shows Mendeleev's ambiguity/ambivalence towards the atomic theory and will be dealt with later (see Step 6).

Step 3: Another example of Mendeleev's ambivalence can be observed from the following: "I shall not form any hypotheses, either here or further on, to explain the nature of the periodic law; for, first of all, the law itself is too simple*" (Mendeleev, 1879, p. 292. The asterisk leads the reader to the following footnote: "However, I do not ignore that to completely understand a subject we should possess, independently of observations (and experiences) and of laws (as well as systems), the meanings of both one and the other.")

Step 4: Although Mendeleev stated in 1879 that he would not formulate a hypothesis, 10 years later in his famous Faraday lecture, Mendeleev (1889) not only attributed the success of the periodic law to Cannizaro's ideas on the atomic theory (pp. 636–637), but went on to explicitly formulate the following hypothesis: "[T] he veil which conceals the true conception of mass, it nevertheless indicated that the explanation of that conception must be searched for in the masses of atoms; the more so, *as all masses are nothing but aggregations, or additions, of chemical atoms*" (Mendeleev, 1889, p. 640, emphasis added).

Step 5: Again, at the Faraday lecture, Mendeleev (1889) took extreme care to explain the periodicity of properties of chemical elements on the basis of atomic theory. I cite at length:

> The periodic law has shown that our chemical individuals [atoms] display a harmonic periodicity of properties, dependent on their masses.... An example will better illustrate this view. The atomic weights –
>
> Ag = 108 Cd = 112 In = 113 Sn = 118 Sb = 120 Te = 125 I = 127
>
> steadily increase, and their increase is accompanied by a modification of many properties which constitutes the essence of the periodic law. Thus, for example, the densities of the above elements decrease steadily, being respectively –
>
> 10.5 8.6 7.4 7.2 6.7 6.4 4.9
>
> while their oxides contain an increasing quantity of oxygen:–
>
> Ag_2O Cd_2O_2 In_2O_3 Sn_2O_4 Sb_2O_5 Te_2O_6 I_2O_7
>
> But to connect by a curve the summits of the ordinates expressing any of these properties would involve the rejection of Dalton's law of multiple proportions. Not only are there no intermediate elements between silver, which gives AgCl, and cadmium which gives $CdCl_2$, but, according to the very essence of the periodic law there can be none; in fact a uniform curve would be inapplicable in such a case, as it would lead us to expect elements possessed of special properties at any point of the curve. (Mendeleev, 1889, pp. 640–641)

This is a clear acknowledgment of the role played by the atomic theory to explain periodicity in the periodic table. Furthermore, Dalton's law of multiple proportions is in itself considered as evidence to corroborate the atomic theory: "The discovery of the law of simple multiple proportions was the first great success of Dalton's atomic theory. This law was not induced from experimental results, but was derived from the theory, and then tested by experiments" (Pauling, 1964, p. 26). It is important to note that Mendeleev clearly conceptualized the relationship between the discontinuous function of the periodic properties and its dependence on the law of multiple proportions, which in the ultimate analysis meant atomic theory. To support this claim I once again quote from Mendeleev's Faraday lecture: "[T]he periodic law has clearly shown that the masses of the atoms increase abruptly, by steps, which are clearly connected in some way with Dalton's law of multiple proportions.... While connecting by new bonds the theory of the chemical elements with *Daltons's theory of multiple proportions, or atomic structure of bodies*, the periodic law opened for natural philosophy a new and wide field for speculation" (Mendeleev, 1889, p. 642, emphasis added). Interestingly, Mendeleev even seems to be considering the law of multiple proportions synonymous with Dalton's atomic theory.

Step 6: At this stage I would like to refer to Mendeleev's ambiguity/ambivalence toward the atomic theory. Throughout the nineteenth century positivism was the dominant philosophy, which led all scientific work to be based strictly on experimental observations and all hypothetical propositions were considered speculative and hence nonscientific

(Brush, 1976; Gavroglu, 2000; Holton, 1992). Mendeleev was clearly aware of this and on many occasions went out of his way to emphasize that the periodic "law itself was a legitimate induction from the verified facts" (Mendeleev, 1889, p. 639). In the Faraday lecture, Mendeleev emphasized the inductive aspect of the periodic law in the light of the antiatomist Marcellin Berthelot's (1827–1907) criticism:

> [T]he illustrious Berthelot, in his work *Les origines de l'alchimie*, 1885, 313, has simply mixed up the fundamental idea of the law of periodicity with the ideas of Prout, the alchemists, and Democritus about primary matter. But the periodic law, based as it is on the *solid and wholesome ground of experimental research*, has been evolved independently of any conception as to the nature of the elements. (Mendeleev, 1889, p. 644, emphasis added)

Apparently, Mendeleev's dilemma was that on the one hand he could rightly claim that the periodic law was based on experimental properties of the elements (an aspiration of scientists in the late nineteenth century), and yet he could not give up the bigger challenge, viz., the possible causes of periodicity, and hence the importance of atomic theory.

It would be appropriate to finish this section by the opinion of an exceptional witness (an experimentalist *par excellence*) who has testified in eloquent terms with respect to the positivist intellectual milieu of the late nineteenth century:

> [I]t is of interest to recall that less than 20 years ago there was a revolt by a limited number of scientific men against the domination of the atomic theory in chemistry. The followers of this school considered that the atomic theory should be regarded as a mere hypothesis, which was of necessity unverifiable by direct experiment, and should, therefore, not be employed as a basis of explanation of chemistry. ... This tendency advanced so far that textbooks of chemistry were written in which the word atom or molecule was taboo. (Rutherford, 1915, p. 176)

Role of Predictions in Scientific Theories and Their Implications for the Periodic Table

After having provided evidence for the relationship between periodicity and atomic theory in the development of the periodic table by Mendeleev, in this section I present arguments as to how predictions play an important role in the development of scientific theories. According to Brush (1996), scientists generally propose a hypothesis, deduce its consequences, make predictions, and do experiments to see if the predictions are borne out. Ziman (1978) believes that the fundamental purpose of science is to acquire the means for reliable prediction. Actual scientific practice, however, is much more complex and controversial.

To facilitate understanding, Brush suggests the following types of predictions. (a) Contraprediction: foretelling the existence of unknown elements and their properties. Brush explicitly points out that the discovery of gallium was a contraprediction. (b) Novel prediction: correction of some of the existing atomic weights by Mendeleev (e.g., beryllium changed from 14 to 9, uranium changed from 120 to 240, tellurium changed from 128 to 125). (c) Retrodiction: explanation of a fact known before the theory was proposed. Although, according to Brush, the convincing power increases in the following order: retrodiction, novel prediction, and contraprediction, there is no consensus among philosophers on this issue.

Lakatos (1970) in his Methodology of Scientific Research Programs conceptualizes the role of predictions in the following terms:

Step 1: "The sophisticated falsificationist regards a scientific theory T as falsified if and only if another theory T′ has been proposed with the following characteristics: (1) T' has excess empirical content over T: that is it predicts *novel* facts, that is, facts improbable in the light of, or even forbidden, by T" (p. 116, original emphasis; in a footnote on this page Lakatos adds that he uses "prediction" in a wide sense that includes "postdiction"). It is plausible to suggest that Mendeleev's contribution provided excess empirical content over that of various other discoverers of the periodic table and furthermore made explicit predictions of unknown elements (cf. van Spronsen, 1969).

Step 2: "[A] series of theories is *theoretically progressive* (*or 'constitutes a theoretically progressive problem-shift'*) if each new theory has some excess empirical content over its predecessor, that is, if it predicts some novel, hitherto unexpected fact" (p. 118, original emphasis. In a footnote on this page Lakatos clarifies further: "A *new* fact must be improbable or even impossible in the light of previous knowledge" (original emphasis)). Although Lakatos's methodology refers to a series of theories, he also recognizes that "there is nothing wrong in saying that an isolated, single theory is 'scientific' if it represents an advance on its predecessor, as long as one clearly realizes that in this formulation we appraise the theory as the outcome of – and in the context of – a certain historical development" (p. 119). Mendeleev's contribution certainly complies with this requirement if we consider it as one of the many attempts to understand the properties of the elements (cf. van Spronsen (1969) for a historical development). According to Papineau (1979) the idea of progressiveness in Lakatos's methodology is closely related to predictions: "A programme is *empirically progressive* as well as theoretically progressive in so far as some of the new predictions are actually borne out" (p. 97, original emphasis).

Step 3: "The time-honoured empirical criterion for a satisfactory theory was agreement with the observed facts. Our empirical criterion for a series of theories is that it should produce new facts. *The idea of growth and the concept of empirical character are soldered into one*" (p. 119, original emphasis). This helps to understand the controversy with respect to the role played by accommodation (agreement of observed facts with the theory) and prediction of new elements by Mendeleev (Brush, 2005; Lipton, 2005). Following Lakatos, it appears that both accommodations and predictions are equally important for progress in scientific theories.

Step 4: Lakatos asks a very pertinent question: "*how are research programmes eliminated*?" (p. 155, original emphasis), and responds by suggesting "by a rival research programme which explains the previous success of its rival and supersedes it by a further display of *heuristic power*" (p. 155, original emphasis). In a footnote Lakatos explains, "I use '*heuristic power*' … to characterize the power of a research programme to anticipate theoretically novel facts in its growth. I could of course use '*explanatory power*'" (p. 155, original emphasis). This clearly shows the role played by the prediction of novel facts based on the negative heuristic of a research program in scientific theories.

Step 5: Finally, Lakatos differentiates between mature and immature science: "*Mature science consists of research programmes in which not only novel facts but, in an important sense, also novel auxiliary theories, are anticipated; mature science – unlike pedestrian trial-and-error – has 'heuristic power'*" (p. 175, original emphasis). Anticipation of novel facts thus plays a crucial role in Lakatos's methodology.

At this stage it is important to note that Lakatos's idea of novel facts based on predictions has been the subject of considerable controversy in the literature (Frankel, 1979; Gardner, 1982; Lakatos & Zahar, 1978; Musgrave, 1974; Nunan, 1984; Thomason, 1992; Zahar, 1973). A detailed discussion goes beyond the subject of this chapter. Murphy (1989) has provided a synthesis and conceptualized the idea of novel facts in the following terms: "A fact that played no role in the formulation of the theory is one whose existence, relevance to the theory, or interpretability in light of the theory was first documented after that theory was first proposed" (p. 388). In my opinion, Mendeleev's prediction of at least three elements (gallium, scandium, and germanium) and their respective properties, do comply to a fair degree with this definition of novel facts. Interestingly, as early as 1877 (about 2 years after the discovery of gallium) a leading British chemist not only recognized Mendeleev's contribution as a theory, but also referred to methodological issues: "The prevision of phenomena not yet observed has been rightly declared by methodologists to be one of the principal distinctions between science, in the strict sense of the term, and a mere accumulation of unorganized knowledge; the discovery of gallium thus shows the value of Mendeleev's theory" (Crookes, 1877, pp. 296, 298). It is important to note that Mendeleev's contribution, in contrast to many of his predecessors, cannot be considered as a mere accumulation of knowledge, but rather has the basic elements of a scientific theory. Weisberg (2007) has argued cogently that Mendeleev had no empirical knowledge with respect to empty slots pertaining to missing elements. Consequently, he first needed to hypothesize the existence of the missing elements based on his theoretical framework and then make predictions. Similarly, van Spronsen (1969), in spite of his ambivalence with respect to Mendeleev's contribution, does recognize the role played by predictions: "After Lecoq de Boisbaudran had discovered gallium in 1875, Mendeleev rightly concluded that the validity of the periodic system of elements could no longer be questioned. The confirmation of this prediction may certainly be called the culminating point in the history of the periodic system" (p. 221).

Relative Importance of Accommodations and Predictions as Possible Support for Mendeleev's Periodic Table

Mendeleev (1869) enunciated the first form of his periodic law and later elaborated in the following terms: "The properties of simple bodies, the constitution of their compounds, as well as the properties of these last, are periodic functions of the atomic weights of elements" (Mendeleev, 1879, p. 267). It is important to note that the elucidation of the concept of atomic weight by Stanislao Cannizaro at Karlsruhe was crucial in the discovery of the periodic law. According to van Spronsen (1959), elaboration of the periodic table was difficult and took a long time due to "lack of a definite conception of atomic weight, which is very closely connected with the definitions of molecules and atoms" (p. 565). Availability of the atomic weights of about 60 elements enabled Mendeleev to accommodate the elements in the table according to various physicochemical properties (density, specific heat, atomic weight, atomic volume, melting point, valence, oxides, chlorides, and sulfides). In contrast to other discoverers, Mendeleev's work was characterized by the following aspects: the division into main and subgroups, the vacant spaces left for undiscovered elements together with the prediction of some of their properties, i.e., the homologues of aluminium and silicon, the classification of the transition metals, and the reversal of tellurium–iodine (van Spronsen, 1969). Historians and philosophers of science continue to debate as to what was crucial for the acceptance of the periodic law by the scientific community: accommodation of the existing elements or the prediction of new ones (Brush, 1996; Dmitriev, 2004; Gordin, 2004; Kaji, 2003; Lipton, 1991; Maher, 1988). Lipton (1991) and Maher (1988) favor a predictivist thesis, viz., Mendeleev's law was accorded a greater recognition after the discovery of the first predicted element (gallium) in 1875. Mendeleev left various vacant spaces in his table and made many predictions and, of these, the following are the most important: (a) eka-aluminum (gallium), (b) eka-boron (scandium), and (c) eka-silicon (germanium). Besides the atomic weights and physical properties (just presented), some of the chemical properties (formation of oxides and chlorides) of the predicted elements coincided to a remarkable degree with the discovered elements. According to van Spronsen (1969), after the discovery of gallium in 1875, Mendeleev rightly concluded that the validity of the periodic system of elements could no longer be questioned. The confirmation of this prediction may certainly be called the culminating point in the history of the periodic system. This precisely is the point of contention among philosophers of science, viz., what made Mendeleev's periodic law valid – accommodations dating from 1869 or the predictions from 1875 onwards.

More recently, there has been considerable debate among historians and philosophers of science with respect to the relative importance of accommodations and predictions and especially in the context of Mendeleev's contribution (Allchin, 2005; Brush, 2005, 2007a; Lipton, 2005a, b). Lipton (2005a) has espoused the predictivist thesis in the following terms: "[T]here is an argument that predictions ought to count more than accommodations, because of the risk of 'fudging' that accommodations run and predictions avoid" (p. 219). Allchin (2005) has endorsed just the opposite, that is, accommodation by arguing that "[p]redictors can fudge too, through excess predictions. They may also hedge by using alternative versions of a hypothesis. Thus, Mendeleev's predictions of eka-aluminium (gallium) and eka-boron (scandium) are widely celebrated, whereas his failed predictions … are largely forgotten" (p. 1409). Brush (2005) is skeptical of the predictivist thesis, but still recognizes the importance of predictions in the case of Mendeleev. More recently, Brush (2007) has clarified his position further:

> I do not argue for or against predictivism in the normative sense that philosophers of science employ, rather I describe how scientists themselves use facts and predictions to support their theories. I find wide variations, and no support for the assumption that scientists use a single "Scientific Method" in deciding whether to accept a proposed new theory. (p. 256)

This makes lot of sense, as scientists can hardly be expected to follow the scientific method, and thus the role of accommodations or predictions would always vary according to the needs of the research program. Now let us see how general chemistry textbooks deal with these issues. Brito et al. (2005) have reported that of the 57 textbooks (all published in the USA) almost all emphasized the importance of accommodations, 30 recognized the importance of predictions, and none the relative importance of accommodations and predictions. This is precisely a major objective of this chapter, namely in order to understand the dynamics of scientific progress, science teachers and textbooks must recognize the importance of both accommodations and predictions.

Mendeleev's Contribution: Theory or an Empirical Law?

Based on evidence provided in the previous sections, here I present arguments as to whether Mendeleev's contribution was a theory or an empirical law. There seems to be considerable controversy among philosophers of science with respect to the nature of Mendeleev's contribution. Wartofsky (1968) clearly considers Mendeleev's contribution to be more than a simple empirical law:

> Mendeleev, for example, predicted that the blank space of atomic number 32, which lies between silicon and tin in the vertical column, would contain an element which was grayish-white, would be unaffected by acids and alkalis, and would give a white oxide when burned in air, and when he predicted also its atomic weight, atomic volume, density and boiling point, *he was using the periodic table as a hypothesis from which predictions could be deduced.* This was in 1871. (p. 203, emphasis added)

Actually, prediction of new elements by Mendeleev was no simple process. The atomic weight and the physicochemical properties had to adjust and follow the pattern dictated by elements preceding and following (in the period), and the elements above and below (in the group). These were stringent preconditions based on the hypothesis that the elements in the periodic table could be arranged in ascending order of their atomic weights provided they comply with the properties of neighboring elements in the periods/groups, and this precisely facilitated the prediction of new elements (novel facts in the Lakatosian framework).

Ziman (1978) recognizes the importance of predictions with respect to the persuasiveness (i.e., heuristic power) of a theory, and hence Mendeleev's contribution can be considered as a theory:

> Needless to say, the most impressive way of validating a scientific theory is to confirm its predictions … the persuasive power of a successful prediction arises from the fact that it could not have been deliberately contrived. The most famous examples, such as Mendel'eef's prediction of the existence and properties of undiscovered elements to fill the gaps in the periodic table, or Gell-Mann and Neeman's prediction of the Omega- minus particle to complete an SU(3) octet, have astonishing rhetorical power. (p. 31, original emphasis)

Shapere (1977) refers to the fact that historically Mendeleev's work has been referred to as a classification, system, table, or law. Nevertheless, in his opinion the periodic table is neither a law nor a theory, but rather an ordered domain: "[A] lthough the periodic table was widely referred to as a law, the general opinion of the time was that it could be called a 'law' only in a rather loose sense, the true law being the precise mathematical expression of the 'function' relating the atomic weights and the other properties of the elements" (p. 536). Interestingly, Gordin (2004, p. 31) has carefully reconstructed Mendeleev's ideas from August 1869 to November 1870 to show how these evolved from considering his contribution a "regularity" to its "lawlike" character. Weisberg (2007) has clarified that if theories allow us to unify, make predictions, and frame explanations, then according to Shapere's own standard, Mendeleev must be considered as a theorist and his contribution as a theory. Similar to Shapere, Bensaude-Vincent (1986) has suggested that Mendeleev

> was able to accomplish the positivist ideal for a mature science: to summarize all the known facts and laws in a systematic table. ... Mendeleev belonged to a strict positivist tradition: his rejection of all hypotheses on the origin of the elements, his search of a single general law gathering the largest number of chemical data, his practice of classification, are all typical attitudes of the "esprit positif" according to A. Comte. (p. 14)

This attribution of "esprit positif" to the work of Mendeleev, however, contrasts sharply with what Mendeleev had to say about his contribution:

> If statements of fact themselves depend upon the person who observes them, how much more distinct is the reflection of the personality of him who gives an account of methods and of philosophical speculations which form the essence of science! For this reason there will inevitably be much that is subjective in every objective exposition of science. (Preface to the 6th Russian ed., *Principles of Chemistry,* Mendeleev, 1897, p. vii)

The first edition of Mendeleev's *Principles of Chemistry* was published in 1869 and was in part an attempt to provide his students information then available through a novel idea (periodic system). According to Gordin (2002) textbooks not only codify knowledge, but also serve as valuable tools for exploring sociological/epistemological construction of a discipline.

This leads one to ask: Was Mendeleev an inductivist or a falsificationist? At this stage it is interesting to refer to difficulties faced by Mendeleev with respect to atomic weight reversals (e.g., nickel and cobalt; tellurium and iodine). In his periodic table, Mendeleev had placed tellurium before iodine, knowing fully well that the atomic weight of tellurium was higher than that of iodine. Despite various efforts to determine the atomic weight of tellurium, it still remained higher than that of iodine. Based on the chemical properties (theoretical framework if you prefer), Mendeleev insisted that tellurium must be placed before iodine. Similarly, the placing of an inert gas (argon) generated lot of controversy, and constituted a crucial test for Mendeleev's theory. This clearly shows how placing of an element was not a straightforward question of ordering the elements in the ascending order of their atomic weights, but rather Mendeleev used a hypothesis from which predictions could be deduced. In other words, his hypothesis was that tellurium must have certain physical and chemical properties from which he could predict its place in the periodic table. As the atomic weight of tellurium then available did not allow it to be placed before iodine, Mendeleev, based on the strength of his hypothesis (theory), insisted that there must be an error in the determination of the atomic weight itself. Now, in order to respond to the question we raised at the beginning of this paragraph, let us seek help from Christie and Christie (2003, p. 173): "Had Mendeleev been an inductivist or a falsificationist, he would surely have had to abandon his scheme! [authors are, of course, referring to Mendeleev having insisted that tellurium be placed before iodine, despite empirical evidence to the contrary]." This clearly shows Mendeleev's *modus operandi*: based on the conviction of his theoretical framework he refused to accept the empirical evidence and insisted on the prediction that followed from his hypothesis.

It appears that historians and philosophers of science generally conceptualize scientific progress to be dichotomous, viz., experimental observations lead to scientific laws, which later facilitate the elaboration of explanatory theories. On the contrary, Lakatos (1970) has argued that "the clash is not 'between theories and facts' but between two high-level theories: between an *interpretative theory* to provide the facts and an *explanatory theory* to explain them; and the interpretative theory may be on quite as high a level as the explanatory theory" (p. 129, original emphasis). In other words, scientific progress is characterized by a series of theories or models (plausible explanations), which vary in the degree to which they explain/interpret/predict the experimental findings.

At this stage I would like to contrast the views of philosophers of science with respect to Mendeleev's work (presented above) and the role of idealization in science. Most scientists and philosophers of science would consider Newton's law of gravitation as a paradigm case of a natural law. Nevertheless, according to Cartwright (1983) for bodies having mass and charge, Newton's law of gravitation and Coulomb's law interact to determine the force between two bodies. She goes beyond and suggests that the two laws are not true and perhaps not even approximately true (see Chapter 2 for more details).

Let us now consider an alternative account from a philosopher of science which provides a way of understanding the practice of science without the laws of nature:

> But one need not appeal to history to deconstruct the concept of a law of nature. The concept is theoretically suspect as well. For example, any law of nature refers to only a few physical quantities. Yet nature contains many quantities which often interact one with another, and there are few if any truly isolated systems. So there cannot be many systems in the real world that exactly satisfy any purported law of nature. Consequently, understood as general claims about the world, most purported laws of nature are in fact false. So we need a portrait of science that captures our everyday understanding of success without invoking laws of nature understood as true, universal generalizations. (Giere, 1995, p. 109)

More recently, Giere (1999) has elaborated further with respect to "science without laws." It is interesting to observe how the conceptualizations of Wartofsky (1968) and Ziman (1978) coincide on the one hand, and the degree to which they differ with those of Bensaude-Vincent (1986) and Shapere (1977) on the other. In contrast, Cartwright (1983), Giere (1995, 1999), Lakatos (1970), and perhaps (!) Mendeleev (1897) conceptualize the problem in an entirely different framework, which in our opinion is quite helpful in understanding Mendeleev's periodic table. In other words, we do not necessarily have to follow the theory/law (or for that matter ordered domain, codification scheme) dichotomy, but rather it is plausible to suggest that Mendeleev's work can be considered as an "interpretative" theory, which became "explanatory" (cf. Lakatos, 1970) after the periodic law was based on atomic numbers (based on Moseley, 1913, 1914; Thomson, Bohr, and Lewis; see next section).

Contributions of Thomson, Lewis, Bohr, and Moseley

It is important to note that various well-known scientists continued a critical appraisal of Mendeleev's efforts in order to provide a better theoretical basis for the periodic table. Thomson (1897), in his celebrated article, had already suggested a possible explanation of the periodic law and later gave a detailed explanation of how properties of the elements in a period varied with respect to the number of corpuscles (electrons): "The gradual change in the properties of the elements which takes place as we travel along one of the horizontal rows in Mendele'efs arrangement of the elements, is also illustrated by the properties possessed by these groups of corpuscles" (Thomson, 1904, p. 259).

Lewis in an unpublished memorandum dated 1902 (reproduced in Lewis, 1923) presented a theory of the cubic atom:

> In the year 1902 (while I was attempting to explain to an elementary class in chemistry some of the ideas involved in the periodic law) becoming interested in the new theory of the electron [Thomson's discovery of the electron in 1897], and combining this idea with those which are implied in the periodic classification, I formed *an idea of the inner structure of the atom* [model of the cubic atom] which, although it contained crudities, I have ever since regarded as representing essentially the arrangement of the electrons in the atom. (Lewis, 1923, pp. 29–30, emphasis added)

In Lewis's model of the cubic atom, a cube reaches its maximum capacity of eight electrons with the last element of the period (noble gas), and then this cube becomes, in some sense, the kernel around which the larger cube of the next period is built. Lewis thought that his model could explain well the formation of polar bonds but not those in the hydrocarbons.

Bohr (1913a, b, c) in his first major publication (the trilogy) tried to establish a relationship between electron configuration and periodicity of the elements:

> In these considerations [electron configurations] we shall assume that the number of electrons in the atom is equal to the number which indicates the position of the corresponding element in the series of elements arranged in order of increasing atomic weight. Exceptions to this rule will be supposed to occur only at such places in the series where deviation from the periodic law of the chemical properties of the elements are observed. (Bohr, 1913b, pp. 486–487)

Moseley (1913, 1914) found that the frequencies of the x-rays given off by cathode-ray tubes depended on the metal used as the anode. He found a linear relationship (included in most textbooks) between the square root of the frequencies of the x-rays and the atomic number, viz., position of the element in the periodic table.

This led him to conclude that the atomic number of an element is equal to the positive charge on the nucleus of an atom. Before determining the high-frequency spectra of the elements, Moseley was fully aware of van den Broek's hypothesis, according to which all chemical and optical properties of an element were determined by its atomic number (Z), that is, by its serial order in the periodic table and not its atomic weight (Heilbron, 1966). Finally, the struggles of so many workers had borne fruit and the scientific community found a plausible cause of the periodicity of the elements as a function of atomic number. According to Heilbron (1966), one of the first post-Moseley periodic tables was built by Ladenburg (1920), which still had vacant spaces for elements 43, 61, 72, and 75. Again, as with most scientific advances, Moseley's contribution was no panacea and soon became the subject of controversy and criticism (cf. Bohr, 1913, 1914; Lindemann, 1913, 1914; Moseley, 1913, 1914).

Mendeleev's Periodic Law: Does It Follow a "Baconian Inductive Ascent"?

In order to understand the Baconian "inductive ascent," Lakatos (1970) has referred to Bohr's model of the atom (see Chapter 1). A major premise of historians who follow the Baconian inductive ascent is that scientific theories and laws are primarily driven by experimental observations. Furthermore, such empiricist interpretations consider scientific progress to be dichotomous, viz., experimental observations lead to scientific laws, which later facilitate the elaboration of explanatory theories. In other words, scientific progress is characterized by a series of theories or models (plausible explanations), which vary in the degree to which they explain/interpret/predict the experimental findings.

It is plausible to suggest that the development of the periodic table can also be conceptualized as a Baconian inductive ascent by scientists, philosophers, and textbook authors with an inductivist perspective, according to the following periods:

1. Early attempts to classify the elements starting from 1817 and the discovery of about 60 elements along with their physical and chemical properties. This period extended up to approximately 1860 (in the case of Bohr's model this corresponds to the chaos of spectrum lines before Balmer's law).
2. Work in this period was stimulated by the Karlsruhe congress of 1860 and important contributions were made by De Chancourtois, Odling, Meyer, Newlands, and Hinrichs (cf. Brito et al., 2005). Mendeleev, of course, has received major credit for having explicitly stated the periodic law in 1869 (this corresponds to Balmer's empirical law for the hydrogen line spectrum).
3. Work of Moseley (1913, 1914) and others finally provided an explanation of the periodic table based on atomic numbers – the modern form of the law, viz., properties of the elements are a periodic function of their atomic numbers (this corresponds to Bohr's explanation of the hydrogen line spectrum).

Why is it so difficult to understand what exactly Mendeleev was doing? Let us recapitulate some of the arguments that I have presented in this chapter. Mendeleev's basic idea was that atomic weights were of fundamental importance in the construction of the periodic table. The next question that we need to address is: Why does a simple arrangement of the elements according to their increasing atomic weight facilitate prediction and accommodation? Mendeleev (1889) himself provided the answer by considering the atomic weights to be nothing but aggregations of chemical atoms. So Mendeleev, based on an idea, hypothesis, or theory, made predictions and this is what many scientists do. Of course, the picture was later refined by Moseley (1913) and many other scientists over the years. This apparently is the cause of controversy. As the more plausible (theoretical, greater heuristic power according to Lakatos, 1970) explanation was provided by Moseley, Mendeleev's contribution could not be considered as an explanation. This line of reasoning goes counter to our understanding that scientific theories are tentative. Based on these arguments, Niaz, Rodríguez, and Brito (2005) have suggested that both Mendeleev's and Moseley's contributions are tentative and they vary in the degree to which they provide explanation for the periodic table. According to Lakatos (1970), theories are superseded in the degree to which a new theory provides greater heuristic power. Actually, this is an important issue, not only for historians and philosophers of science, but also for science students. No one in 1869, 1879, or 1889, or for that matter until 1913, could tell that the atomic number and not the atomic weight was the "true explanation" of the periodic table. It is only with hindsight that we can consider Mendeleev's periodical law to be lacking a "theoretical explanation." Furthermore, how can we know that the atomic number is the final explanation of the periodic table – a more fundamental particle may change this "true basis." In a similar vein, Weisberg (2007) has pointed out that just like any other theory, Mendeleev's periodic system needed further theoretical explanation. Burbules and Linn (1991, p. 232) have argued cogently as to how history of science shows that, in the long run, all theories more or less turn out to be "wrong." The periodic table provides a good opportunity to facilitate students' understanding of the tentative nature of science (Niaz et al., 2004).

Educational Implications of a Historical Reconstruction of the Periodic Table

Previous sections have provided a perspective of the periodic table and its construction based on a history and philosophy of science framework. In this section, some educational implications are explored based on the following heuristic principles (criteria).

Explanation of Periodicity in the Periodic Table

The objective behind this criterion was to make students think and reason with respect to the possible causes of periodicity in the periodic table. Many students must have wondered how a simple arrangement could provide such regularities. Textbooks could promote students' curiosity, and a historical reconstruction of the periodic table provides an opportunity to facilitate this objective by emphasizing either: (a) inductive generalization, and/or (b) periodicity as a function of atomic theory. Brito et al. (2005) have reported that of the 57 general chemistry textbooks (published in the USA) none accomplished this objective satisfactorily, 14 made a simple mention, and 43 simply ignored the issue. It is important to note that even those textbooks that ignored the issue implicitly recognized that the periodic table was a consequence of the accumulation of experimental data. Of the 14 textbooks that made a simple mention, some emphasized inductive generalization, and following are two examples:

> Mendeleev's approach to the periodic table was empirical; he based his classification scheme on the observed facts. (Hill & Petrucci, 1999, p. 316)

> The periodic table was created by Mendeleev to summarize experimental observations. He had no theory or model to explain why all alkaline earths combine with oxygen in a 1:1 atom ratio – they just do. (Moore et al., 2002, p. 266)

In light of the historical reconstruction presented, to state that the periodic table was empirical and that Mendeleev had no theory or model to explain the periodicity of the properties of the elements is perhaps rather simplistic and difficult to sustain. It is more fruitful and plausible to present a more balanced picture to the students by highlighting the dilemma faced by Mendeleev (and others) in which they struggled to look for underlying patterns to explain and understand periodicity.

Importance of Prediction as Evidence to Support the Periodic Law

Brito et al. (2005) have reported that of the 57 general chemistry textbooks analyzed, 30 emphasized the importance of prediction satisfactorily as evidence to support the periodic law and, of these, 29 compared the properties of at least one of the predicted elements (Ga, Sc, and Ge) with the experimental values. This comparison was presented in the form of a table occupying about one half of a page. Most textbooks presented arguments to emphasize the role of predictions; for instance: "It was the extraordinary success of Mendeléeff's predictions that led chemists not only to accept the periodic table but to recognize Mendeléeff more than anyone else as the originator of the concept on which it was based" (Bodner & Pardue, 1989, p. 201). According to Hill and Petrucci (1999): "The *predictive* nature of Mendeleev's periodic table led to its wide acceptance as tremendous scientific accomplishment" (p. 45). One of the textbooks (Phillips et al., 2000) compared the prediction of the elements and their properties to that of Halley's comet, which repeats its cycle every 76 years, and included an exercise in which the students are asked to predict the properties of an unknown element (Ge), while having the properties of Si, Ga, As, and Sn. Twenty-five textbooks reproduced Mendeleev's 1871 periodic table (at times in color and various devices to highlight missing elements), occupying about one half of a page to emphasize the elements predicted.

Relative Importance of Accommodation and Prediction in Development of the Periodic Table

Historically, scientists have strived to explain the periodic table on the basis of predictions and accommodations (placing of the different elements in the periodic table according to their physical and chemical properties). In the previous criterion it was observed that about half of the textbooks analyzed had recognized the importance of predictions (Brito et al., 2005). Surprisingly, however, none of the textbooks explained satisfactorily and only six made a simple mention of the relative importance of accommodations and predictions. This shows that textbooks lack the appreciation that there are alternative interpretations with respect to the success of the periodic table. One textbook came quite close to having a satisfactory presentation:

> Any good hypothesis must do two things: It must explain known facts [accommodation], and it must make predictions about phenomena yet unknown. ... Mendeleev's hypothesis about how known chemical information could be organized passed all tests. Not only did the periodic table arrange data in a useful and consistent way to explain known facts about chemical reactivity, it also led to several remarkable predictions that were later found to be accurate. (McMurry & Fay, 2001, p. 160)

This is a fairly good presentation of Mendeleev's dilemma (hypothesis), and could have been classified satisfactory had the authors recognized the role of controversy and alternative interpretations. The inclusion of alternative interpretations could facilitate students' understanding of how progress in science inevitably leads to controversies and rivalries and, at times, it is difficult to foresee and predict all implications of a theory.

Mendeleev's Contribution: Theory or Empirical Law?

This criterion is essential in understanding the nature of Mendeleev's contribution, viz., what exactly was he trying to do with all the information available. Mendeleev's own ambivalence notwithstanding, the historical reconstruction shows that Mendeleev's ingenuity consisted of precisely recognizing not only that the periodic table was a "legitimate induction from the verified facts" but also that there was a reason/cause/explanation for this periodicity, viz., the atomic theory. In other words, scientists do not decide beforehand that their contribution would be empirical/theoretical, but rather the scientific endeavor inevitably leads them to "speculate" with respect to underlying patterns of what they observe. Mendeleev's case is an eloquent example of this dilemma. None of the textbooks made a satisfactory presentation and 52 simply ignored the issue (Brito et al., 2005). Only five textbooks made a simple mention and, of these, Lippincott et al. (1977) considered the periodic table to be an ordered domain: "If we examine the nature of scientific studies, we find that they always start with a group of observations collected as the data available for contemplation. The second step in the study is that of classification of data into recognizable related groupings. … The periodic table … is an example of descriptive classification and ordering" (pp. 304–305). Stoker (1990) considered Mendeleev's contribution to be an empirical law:

> For many years after the formulation of the periodic law and the periodic table, both were considered to be empirical. The law worked and the table was very useful, but there was no explanation available for the law or for why the periodic table had the shape it had. It is now known that the theoretical basis for both the periodic law and the periodic table lies in electronic theory. (p. 155)

This presentation is quite representative of most textbooks. It ignores the fact that scientists were constantly trying to look for a "theoretical basis" of the periodic table, including Mendeleev himself. However, to state that for many years the periodic table had no explanation is to ignore that progress in science is always tentative. In other words, our theories can hardly be considered to be final – in the future we may find a better explanation of the periodic table than that provided by the electronic theory. McMurry and Fay (2001) provide an example of how Mendeleev's contribution can be considered a theory:

> In many ways, the creation of the periodic table by Dmitri Mendeleev in 1869 is an ideal example of how a scientific theory comes into being. At first, there is only random information – a large number of elements and many observations about their properties and behavior. As more and more facts become known, people try to organize the data in ways that make sense, until ultimately a consistent hypothesis emerges. (p. 160)

As we have observed in this chapter the development of the periodic table is much more complex. Nevertheless, recognition of the role played by "emerging hypotheses" can facilitate a better understanding of the vicissitudes faced by Mendeleev and others, in their struggle to go beyond the observable entities.

Textbooks give the impression that for almost 100 years (1820–1920) scientists had no idea or never asked the question as to whether there could be an underlying pattern to explain periodicity. In other words, textbooks provide students a noncontroversial "finished product" that could explain periodicity and the nature of the periodic table only when the modern atomic theory was formulated. The textbook approach does not facilitate students' understanding with respect to the tentative nature of science, considered to be important by modern philosophers of science and also science educators (for details, see Chapter 2). Furthermore, neither science nor scientists can provide the final/true explanation.

Periodic Table as a "Baconian Inductive Ascent"

Following Lakatos (1970), a Baconian inductive ascent would incorporate the following sequence of periods: (a) accumulation of data with respect to the elements and early attempts at classification, starting in 1817; (b) postulation of Mendeleev's periodic law in 1869, as an inductive generalization (based on atomic mass); and (c) explanation of the periodic law based on the work of Moseley (1913) and the electronic theory (based on atomic number). Most textbooks explicitly refer to one or two of the periods (cf. Brito et al., 2005). The following six textbooks explicitly refer to the three periods and their presentation can be considered a "Baconian inductive ascent": Brady et al. (2000), Hill and Petrucci (1999), Kotz and Purcell (1991), Lippincott et al. (1977), Sisler et al. (1980), and Umland (1993). Three examples are provided here.

Brady et al. (2000)

a) The need for organization was recognized by many early chemists, and there were numerous attempts to discover relationships among the chemical and physical properties of the elements. (p. 61)
b) On the basis of extensive observations of this type, Mendeleev devised the original form of the **periodic law**. (p. 62, emphasis in original)
c) Apparently, then, atomic mass was not the true basis for the periodic repetition of the properties of the elements. To determine *what the true basis was*, however, scientists had to await the discoveries of the atomic nucleus, the proton, and atomic numbers. (p. 63, emphasis added)

Hill and Petrucci (1999)

a) In the nineteenth century, chemists discovered elements at a rapid rate. By 1830, 55 elements were known, all with different properties and no apparent pattern among them. (p. 45)
b) Mendeleev's approach to the periodic table was *empirical*; he based his classification scheme on the observed facts. (p. 316, original italics)
c) [W]e will learn that Mendeleev's scheme also makes sense from a theoretical standpoint. We will first learn how electrons are distributed among the regions of an atom described by atomic orbitals, a description called the *electron configuration* of the atom. Then we will explore electron configurations as a basis for the periodic table. (p. 316, original italics)

Kotz and Purcell (1991)

a) In the 19th century many chemists tried to find relationships between atomic weights and the properties of the elements. These efforts largely failed because atomic weights were not known for all the elements, and many measured values were inaccurate. (p. 334)
b) In spite of Mendeleev's great achievement, problems arose when new elements were discovered and more accurate atomic weights determined. ... The *fault lies with Mendeleev's assumption* that the properties of the elements are periodic functions of their atomic weight. (p. 334, emphasis added)
c) Indeed, if the elements are arranged in order of increasing atomic number, *the defects in the Mendeleev table are corrected.* It was therefore Moseley who discovered the **law of chemical periodicity**: the properties of the elements are periodic functions of their atomic numbers. (p. 334, emphasis in the original and italics added)

These presentations are problematic for various reasons, and it is worth reproducing at length, as they provide greater insight with respect to the tentative nature of science:

(a) no scientist can be aware/foretell that his contribution would constitute the "true basis" of observed phenomena; (b) Mendeleev's faults and defects were not considered so by his contemporaries; (c) no one in 1869, 1879, or 1889, or for that matter until 1913, could tell that the atomic number and not the atomic weight was the "true basis" of the periodic table; (d) it is only with hindsight that we can refer to the true basis, faults, and defects in Mendeleev's contribution; and (e) how can we know that the atomic number is the final explanation of the periodic table – more fundamental particles (perhaps quarks!) may change this true basis. It is plausible to suggest that textbooks can provide students with a better appreciation of scientific progress by emphasizing the tentative nature of science. Burbules and Linn (1991, p. 232) explain cogently that if the history of science shows anything, it is precisely that, in the long run, all theories more or less turn out to be "wrong."

This chapter has shown that despite Mendeleev's own ambivalence, periodicity of properties of chemical elements in the periodic table can be attributed to atomic theory. It is argued that based on the Lakatosian framework, predictions (novel facts) play an important role in the development of scientific theories. In this context, Mendeleev's predictions (along with accommodations) played a crucial role in the development of the periodic table. In order to understand progress in science it is important to emphasize the relative importance of both accommodations and predictions. Brush (2007) has endorsed precisely such a position, namely neither scientists nor philosophers of science recommend a "scientific method" based on accommodations/predictions. Perhaps a more "ecumenical" view would endorse both. Finally, it is concluded that Mendeleev's contribution can be considered as an "interpretative" theory, which became "explanatory" (Lakatos, 1970) after the periodic table was based on atomic numbers.

Chapter 6
Foundations of Modern Atomic Theory: Thomson, Rutherford, and Bohr

Introduction

According to Schwab (1974) scientific inquiry tends to look for patterns of change and relationships which constitute the heuristic (explanatory) principles of our knowledge. In other words: "A fresh line of scientific research has its origins not in objective facts alone, but in a conception, a deliberate construction of the mind … this conception [heuristic principle] … tells us what facts to look for in the research. It tells us what meaning to assign these facts" (Schwab, 1974, p. 164). Monk and Osborne (1997) pointed out how many science curricula have forgotten Schwab's important epistemological distinction between the methodological (experimental data) and interpretative (heuristic principles) components. Matthews (1994) emphasized the importance of heuristic principles in scientific inquiry and science education in similar terms. To understand the function of "heuristic principles" let us consider J.J. Thomson's experimental work with cathode rays. Although the experimental details are important we cannot ignore the rationale behind Thomson's determination of the charge to mass ratio of cathode rays. This rationale, which helped Thomson to identify cathode rays as ions or universal charged particles (rival hypotheses), precisely constitutes the "heuristic principle." In a recent study, Blanco and Niaz (1997b) have shown how both students and teachers understand Thomson's experiments as a series of conclusions based on empirical findings (truths). In the case of Bohr's research program, Lakatos (1970) considers Bohr's explanation of the paradoxical stability of the Rutherford atom as the heuristic principle. In contrast, most textbooks consider Bohr's major contribution to be the explanation of the Balmer and Paschen series of the hydrogen line spectrum (i.e., experimental findings). This reminds us that almost 45 years ago we ignored Schwab's (1962) advice that science cannot be taught as an "unmitigated *rhetoric of conclusions* in which the current and temporary constructions of scientific knowledge are conveyed as empirical, literal, and irrevocable truths" (p. 24, emphasis in original).

The history of the structure of the atom since the late nineteenth and early twentieth centuries shows that the models of J.J. Thomson, E. Rutherford, and N. Bohr evolved in quick succession and had to contend with competing models based on rival research programs. This period of the history of structure of the atom has been the

M. Niaz, *Critical Appraisal of Physical Science as a Human Enterprise: Dynamics of Scientific Progress*, Science and Technology Education Library 36,

subject of considerable debate and controversy in the history and philosophy of science literature (Achinstein, 1991; Falconer, 1987; Heilbron, 1985; Heilbron & Kuhn, 1969; Hettema, 1995; Holton, 1986, 1993; Jammer, 1966; Kuhn, 1984; Lakatos, 1970; Popper, 1965).

According to Schwab (1974), it is important to understand not only the experimental details but also the heuristic principle that underlies the experimental findings:

> In physics, similarly, we did not know from the beginning that the properties of particles of matter are fundamental and determine the behavior of these particles, their relations to one another. *It was not verified knowledge but a heuristic principle, needed to structure inquiry, that led us to investigate mass and charge and, later, spin.* (Schwab, 1974, p. 165, emphasis added)

A historical reconstruction of the foundations of modern atomic theory is presented in the following sections:

1. Thomson's cathode ray experiments
2. Rutherford's alpha particle experiments
3. Bohr's model of the atom

Thomson's Cathode Ray Experiments

Cathode rays were first discovered by Plucker in 1858 (Falconer, 1987), long before Thomson became interested in them. Thomson's early views on the nature of electricity were within the accepted tradition of Maxwell's electrodynamics. The Maxwellian view of electricity was a strained state of the ether and discharge in the cathode-ray tube was a relaxation of this strained state, with a consequent dissipation of energy. It is important to note that when Thomson conducted his experiments he was well aware of the controversy with regard to the nature of cathode rays: Were they particles or were they waves in the ether? (Achinstein, 1991, pp. 299–300; Falconer, 1987, p. 243). The controversy actually began in 1879 with Crookes' (1879) support for a particle theory of cathode rays. Deflection of cathode rays by a magnetic field was considered to provide strong support for a particle theory. Hertz (1883), on the other hand, showed that cathode rays were not deflected by an electrostatic field, contrary to the predictions of the particle theory (Falconer, 1987, p. 244). This finding provided support for the ether theory of cathode rays, according to which they were some sort of ethereal disturbance similar to light. Further support for the ether theory was provided by Goldstein (1880), Weidemann (1884), and Hertz (1892). Thomson's own thinking on the nature of cathode rays seems to have been ambivalent. As early as 1881, he (Thomson, 1881) seems to have conceived cathode rays as charged atoms, rather than charged molecules as Crookes had suggested earlier. Based on an unpublished draft of Thomson's book (*Notes on Recent Researches in Magnetism and Electricity*, Oxford, published in 1893), Falconer (1987, p. 247) concluded that Thomson was rather sympathetic to the ether theory of cathode rays. Surprisingly, however, as late as 1909 Thomson had this to say at a scientific meeting: "The ether is not a fantastic creation of the speculative philosopher; it is as essential to us as the air we breathe. ... The study of this all-pervading substance is perhaps the most fascinating and important duty of the physicists" (Thomson, 1909).

This perhaps illustrates the importance of prior knowledge (alternative conceptions) for both scientists and students. Falconer (1987) further suggests that the controversy about the nature of cathode rays (German physicists generally supporting the ether theory, and the British supporting the particle theory) was not important to Thomson until 1895. It was the discovery of x-rays in 1895 that triggered Thomson's and other physicists' interest in cathode rays. Interestingly, the number of publications referring to the nature of cathode rays increased abruptly from 4 in 1895 to 20 in 1896 (Falconer, 1987, p. 246).

Thomson's (1897) Article in the Philosophical Magazine

Given this new interest in the nature of cathode rays Thomson conducted a series of experiments at the beginning of 1897, which were first presented at a Friday evening discourse of the Royal Institution on April 29, 1897. An abstract was published in *The Electrician* (vol. 39, pp. 104–108) on May 21, 1897, and finally published at length in the *Philosophical Magazine* in October 1897 (Thomson, 1897). It is important to note the following important aspects of Thomson's now famous article:

1. In the very first sentence Thomson states the objective of the experiments, namely, to gain some information as to the nature of the cathode rays (p. 293).
2. In the second sentence he refers to the controversy with regard to the nature of cathode rays: "The most diverse opinions are held as to these rays; according to the almost unanimous opinion of German physicists they are due to some process in the aether ... another view of these rays is that, so far from being wholly aetherial, they are wholly material" (p. 293). At this stage, Thomson seems to favor the particle theory (the article, of course, having been written after the experiments).
3. Thomson explains (deconstructs, according to Falconer [1987], p. 245) why Hertz (1883) could not obtain a deflection of the cathode rays electrostatically, and that it could only be obtained when the vacuum was a good one:

> Hertz made the rays travel between two parallel plates of metal placed inside the discharge-tube, but found that they were not deflected when the plates were connected with a battery of storage cells; on repeating this experiment I at first got the same result, but subsequent experiments showed that the absence of deflexion is due to the conductivity conferred on the rarefied gas by the cathode rays. On measuring this conductivity it was found that it diminished very rapidly as the exhaustion increased; it seemed then that on trying Hertz's experiment at very high exhaustions there might be a chance of detecting the deflexion of the cathode rays by an electrostatic force. (p. 296)

4. Thomson summarizes the properties of the cathode rays (found in most textbooks) and points out a fundamental aspect of his experiments; namely, the cathode rays are the same whatever the gas through which the discharge passes and concludes:

> As the cathode rays carry a charge of negative electricity, are deflected by an electrostatic force as if they were negatively electrified, and are acted on by a magnetic force in just the way in which this force would act on a negatively electrified body moving along the path of these rays, I can see no escape from the conclusion that they are charges of negative electricity carried by particles of matter. The question next arises, What are these particles? are they atoms, or molecules, or matter in a still finer state of subdivision? To throw some light on this point, I have made a series of measurements of the ratio of the mass of these particles to the charge carried by it. To determine this quantity, I have used two independent methods. (p. 302)

This perhaps is the most important aspect of Thomson's article, and shows clearly that he visualized that the determination of the mass (m) to charge (e) ratio (m/e) of the cathode rays would help him to identify the cathode ray particles as ions or a universal charged particle.

5. Thomson reports the results of the mass to charge (m/e) ratio of the cathode ray particles:

> [T]he value of m/e is independent of the nature of the gas, and its value 10^{-7} is very small compared with the value 10^{-4}, which is the smallest value of this quantity previously known, and which is the value for the hydrogen ion in electrolysis. (p. 310)

Thomson goes on to speculate that the smallness of m/e may be due to the smallness of m or the largeness of e, or to a combination of both (p. 310). A little later in the article Thomson suggests that the smallness of m/e was due to both (p. 312).

6. In another important passage in the article, Thomson shares his thoughts with the reader:

> If, in the very intense electric field in the neighbourhood of the cathode, the molecules of the gas are dissociated and are split up, not into the ordinary chemical atoms, but into these primordial atoms, which we shall for brevity call corpuscles; and if these corpuscles are charged with electricity and projected from the cathode by the electric field, they would behave exactly like cathode rays. They would evidently give a value of m/e which is independent of the nature of the gas and its pressure, for the carriers are the same whatever the gas may be. (p. 311)

7. Thomson, of course, not only speculates and shares his thoughts with the reader but also suggests an explanation:

> The explanation which seems to me to account in the most simple and straightforward manner for the facts is founded on a view of the constitution of the chemical elements which has been favourably entertained by many chemists: *this view is that the atoms of the different chemical elements are different aggregations of atoms of the same kind.* In the form in which this hypothesis was enunciated by Prout, the atoms of the different elements were hydrogen atoms; in this precise form the hypothesis is not tenable, but if we substitute for hydrogen some unknown primordial substance X, there is nothing known which is inconsistent with this hypothesis, which is one that has been recently supported by Sir Norman Lockyer for reasons derived from the study of the stellar spectra. (p. 311, emphasis added)

Apparently, Thomson was pursuing various objectives at the same time. On the one hand, he disagreed with Prout's hypothesis, and on the other, he wanted to formulate a new hypothesis (underlined part with atoms substituted by the primordial substance X). However, Thomson did not have any conclusive evidence for his new plausible hypothesis and hence he sought two allies, the chemists and Norman Lockyer. Lockyer (1881) had advanced a theory of a divisable atom to explain the different stellar spectra. Interestingly, Lockyer (1897) had published his new theory in March 1897 just 1 month before Thomson proposed his corpuscle hypothesis. It is plausible to suggest that Thomson's hypothesis could be considered as the negative heuristic, hard core (Lakatos, 1970, p. 133), of his research program. According to Lakatos (1970) the hard core is irrefutable by the methodological decision of the protagonists – which, in Thomson's case, would perhaps be the chemists and Lockyer.

8. Thomson takes his hypothesis to yet another theoretical level by proposing two alternative models of the "chemical atom":

> Hypothesis: If we regard the chemical atom as an aggregation of a number of primordial atoms, the problem of finding the configurations of stable equilibrium for a number of equal particles acting on each other ... can be explained by two models:
>
> Model A: Law of force, for example, like that of Boscovitch, where the force between them is a repulsion when they are separated by less than a certain critical distance, and an attraction when they are separated by a greater distance.
>
> Model B: The simpler model of a number of mutually repellent particles held together by a central force. (p. 313)

9. As a grand finale, Thomson presents a theory of atomic structure:

> Thus on this view we have in the cathode rays matter in a new state, a state in which the subdivision of matter is carried very much further than in the ordinary gaseous state: a state in which all matter – that is, matter derived from different sources such as hydrogen, oxygen, etc. – is of one and the same kind; this being the substance from which all the chemical elements are built up. (p. 312)

Heuristic Principles of Thomson's Model of the Atom

Scrutinizing Thomson's 1897 article it can be observed that he goes far beyond a simple presentation of experimental results by speculating, hypothesizing, proposing models, offering explanations, and formulating a theory, which contrasts with the traditional view of the scientific method (cf. Niaz, 1994). This is what Schwab (1974) refers to as "[m]ost of us were taught a schoolbook version of a syntax under the guise of 'scientific method' " (p. 172). Let us now look at how some of the other experimental physicists received Thomson's article at that time. Interestingly, FitzGerald (1897) proposed an alternative explanation for cathode rays based on "free electrons" in the same issue of *The Electrician* (May 21, 1897) in which an abstract of Thomson's article had appeared prior to publication in the *Philosophical Magazine* (October 1897). Apparently, FitzGerald accepted Thomson's hypothesis that cathode rays were corpuscles/primordial atoms/free electrons, but he questioned (precisely the "hard core") that these corpuscles were constituent parts of all atoms:

> This [FitzGerald's explanation] is somewhat like Prof. J. J. Thomson's hypothesis, except that it does not assume the electron to be a constituent part of an atom, nor that we are dissociating atoms, nor consequently that we are on the track of the alchemists. (FitzGerald, 1897, p. 104)

Determination of the mass to charge ratio (*m/e*) of the cathode rays can perhaps be considered the most important experimental contribution (based on a heuristic principle) of Thomson's 1897 article. Yet, he was neither the first to do so nor the only experimental physicist. Schuster (1890) was perhaps the first to report (*m/e*) ratios for cathode rays, and his value came close to that of a charged nitrogen atom, which led him to conclude that cathode rays were charged atoms. Two German physicists, Kaufmann (1897) and Wiechert (1897), also determined (*m/e*) ratios of cathode rays in the same year as Thomson and their measurements agreed with each other. Falconer (1987) explained cogently how Thomson's contribution differed from that of Kaufmann and Wiechert:

> Kaufmann, an ether theorist, was unable to make anything of his results. Wiechert, while realizing that cathode ray particles were extremely small and universal, lacked Thomson's tendency to speculation. He could not make the bold, unsubstantiated leap, to the idea that particles were constituents of atoms. Thus, while his work might have resolved the cathode ray controversy, he did not 'discover the electron'. (p. 251)

Apparently, Thomson's ability to speculate, elaborate alternative hypotheses and models, and perhaps most importantly formulate a theoretical framework for his experimental findings, led him to foresee and conceptualize what his contemporaries ignored. Thomson's interpretations have been the subject of criticism in the philosophy of science literature. Heilbron (1964), for example, claimed that Thomson's arguments were faulty as he claimed far more for the "corpuscle" than the data authorized and, furthermore that few physicists in 1897 were prepared to believe, on this basis, that the world was made of corpuscles. Achinstein (1991), on the other hand, evaluated Thomson's theorization more favorably (p. 296). Thomson, although no philosopher of science, perhaps tried to respond to his critics in 1907 in the following terms:

> From the point of view of the physicist, a theory of matter is a policy rather than a creed; its object is to connect or coordinate apparently diverse phenomena, and above all to suggest, stimulate and direct experiment. It ought to furnish a compass which, if followed, will lead the observer further and further into previously unexplored regions. (Thomson, 1907, p. 1)

Leon Cooper (Nobel Laureate, physics, 1972) has summarized the reconstruction of Thomson's model of the atom in the following terms: "Not the first time that the greatest difficulty in doing a crucial experiment lay as much in developing necessary techniques as in its conception" (Cooper, 1970, p. 312; also Cooper, 1992, p. 312).

Educational Implications of the Thomson Model of the Atom

At this stage it would be interesting to explore the educational implications of Thomson's experiments and his model of the atom. Results from one study are presented that are quite representative of students' epistemological beliefs. Blanco and Niaz (1998) asked freshman students (18–20-year-olds) registered in a general chemistry course the following question: How would you have interpreted, if on using different gases in the cathode-ray tube, the relation charge to mass (*m/e*) would have resulted different? These students had already studied in their course that J.J. Thomson (1897) had performed experiments to demonstrate that on using different gases in the cathode-ray tube, the relation (*m/e*) remained constant, which led him to postulate the "electron." This question was designed to evaluate students' understanding of a hypothetical experimental finding, which would have led to the postulation of a different atomic model. The following responses are quite illustrative of students' thinking: (a) this was possible only if there would have been a modification in the deflection of the cathode rays on the application of the electric field and other factors; (b) the relation *m/e* could not have been a constant, as it was dependent on the type of gas used in the tube; (c) only if the substances used in the tube were not gases; (d) it could not have been different as Thomson reached his conclusion by varying the gases in the tube and found that all atoms contained electrons; (e) if that would have been the case, Thomson could not have discovered the relation *m/e*; (f) that is a poor interpretation, as we already know that the relation *m/e* is independent of the gas used; (g) this could be attributed to inadequate techniques that would alter the structure and composition of matter or even due to calculation errors. These responses make interesting reading and at best could be considered as ambiguous. However, responses (c) to (g) clearly show that students resist the idea of changing models/tentativeness of science, even when explicitly asked to consider a different experimental situation that would lead to the "construction" of an alternative model. Most of these responses consider Thomson's model as an absolute reality, a law or even perhaps predetermined and hence could not be changed. It is not far-fetched to suggest that such an understanding is at least partially a consequence of the science curriculum and textbooks that emphasize Kuhn's (1970) "normal science" and ignore conflicts, controversies, and presuppositions involved in interpreting experimental data. Furthermore, this raises an important issue: Do we want our students to understand science as a product of the work of geniuses who know beforehand what they are going to discover?

In a follow-up study (Niaz et al., 2002) based on Thomson's experiments (in contrast to most courses and textbooks) freshman students were provided not only the experimental details, but also the opportunity to think, argue, reflect, and discuss the following issues (based partially on the historical reconstruction presented in Niaz [1998], and Blanco and Niaz [1998]): what was most important in Thomson's experiments; on using different gases, why did the charge to mass (*m/e*) relation remain constant; did Thomson's model represent the information he had at that time; why did Thomson determine the *m/e* relation; and finally the opportunity to interpret a hypothetical experimental finding, viz., on using different gases in the cathode-ray tube, the relation *m/e* resulted different. These and other issues were discussed as part of intact classroom activities over a period of about 8 weeks and facilitated greater conceptual understanding of the experimental as compared to control group.

Niaz (1998) has reported that of the 23 general chemistry textbooks (published in the USA) only two described satisfactorily that Thomson determined the ratio (*m/e*) to decide whether cathode rays were ions or a universal charged particle. Following is an example of a satisfactory presentation from a general chemistry textbook:

> A very striking and important observation made by Thomson is that the *e/m* ratio does not depend on the gas inside the tube or the metal used for the cathode or anode. The fact that the *e/m* ratio is the same whatever gas is present in the tube proves that the cathode ray does not consist of gaseous ions, for it did, *e/m* would depend on the nature of the gas. (Segal, 1989, p. 412)

Let us now contrast this with the presentation of a textbook that made no mention of the heuristic principle:

> A physicist in England named J.J. Thomson showed in the late 1890s that the atoms of any element can be made to emit tiny negative particles. (He knew they had a negative charge because he could show that they were repelled by the negative part of an electric field). Thus he concluded that all types of atoms must contain these negative particles, which are now called *electrons*. (Zumdahl, 1990, p. 97, emphasis in original)

This presentation makes no attempt to present arguments, reasons, or strategies used by Thomson in order to postulate his model of the atom, and Schwab (1962) very aptly refers to as a "rhetoric of conclusions."

Rodríguez and Niaz (2004a) have shown that of the 41 general physics textbooks evaluated, only one (Eisberg, 1973) made a simple mention and one (Cooper, 1970) made a satisfactory presentation. Cooper (1970) presents a detailed historical reconstruction based on the original writings of Thomson (1897) that follows the sequence: historical context – Thomson's understanding of the controversy; experimental details – Thomson's strategy to tackle the dilemma; and mathematical details – postulation of the electron. The protocol used for analyzing physics textbooks had eight criteria and was used previously by Niaz (1998) for general chemistry textbooks. One of the most important features of Cooper's (1970) presentation was the extensive use of original historical sources. Of the 41 textbooks analyzed by Rodríguez and Niaz (2004), Cooper (1970) had the highest score of 13 points (out of a maximum of 16), and the second highest score was 4 points. Actually, these quantitative details do not do justice to the well-thought-out history and philosophy of science approach of Cooper (1970). Finally, Cooper (1970) summarizes his reconstruction in the following terms: "Not the first time that the greatest difficulty in doing a crucial experiment lay as much in developing necessary techniques as in its conception" (p. 312), which can provide guidelines for future textbooks.

Rutherford's Alpha Particle Experiments

Before turning his attention to the structure of the atom, Rutherford's (1904, 1906, 1913) main research interest was radioactivity. In their efforts to characterize the nature of the alpha particle, Rutherford and colleagues (Geiger and Marsden) had observed something entirely unexpected and troublesome – "scattering" of alpha particles – the deflection of alpha particles from their true line of flight as they passed through matter. In a letter to Baumbach (Rutherford's glass-blower), written in the summer of 1908, Rutherford complained: "The scattering is the devil" (reproduced in Wilson, 1983, p. 286). However:

> In 1908 the scattering was a technical problem to be overcome – but, as with so many other of Rutherford's great leaps of scientific imagination, when the experiment was over he asked Geiger to look into scattering as a phenomenon in its own right. And from this fascination with the small anomaly great results were to be achieved. (Wilson, 1983, p. 287)

Later in life Rutherford was fond of recounting how Geiger had suggested that the young Marsden could perhaps start working on a small research project. Rutherford's response to Geiger was indeed prophetic: "Why not let him see if any alpha particles can be scattered through a large angle?" (reproduced in Wilson, 1983, p. 291). It did not take long for Geiger and Marsden (1909) to report that beta particles bouncing back off a metal plate was a well-known phenomenon, but their findings were extraordinary: "A small fraction of alpha particles falling upon a metal have their directions changed to such an extent that they emerge again at the side of incidence" (p. 495). Further experiments confirmed the well-known phenomenon of the deflection of alpha particles, described in most textbooks. The birth of the nuclear atom itself can perhaps be traced to a dinner conversation at Rutherford's residence, just before Christmas, 1910, as "after supper the nuclear theory came out" (Darwin, 1962).

Rutherford announced his hypothesis of the nuclear atom for the first time on March 7, 1911, at a meeting of the Manchester Literary and Philosophical Society, which was later submitted in April and published in the *Philosophical Magazine*, May 1911 (Rutherford, 1911).

Rutherford's (1911) Article in the Philosophical Magazine

1. In the very first paragraph, Rutherford starts on a controversial note by pointing out that

> [i]t has generally been supposed that the scattering of a pencil of alpha or beta rays in passing through a thin plate of matter is the result of a multitude of small scatterings by the atoms of matter traversed. (p. 669)

This of course referred to the experimental work of Crowther (Proceedings of the Royal Society, lxxxiv, 1910, p. 226), a colleague of Thomson. Based on Crowther's work, Thomson propounded the *compound scattering hypothesis* as a rival to the *single scattering hypothesis* of Rutherford. Rutherford goes on to explain briefly that Thomson's model of the atom is supposed to "consist of a number N of negatively charged corpuscles, accompanied by an equal quantity of positive electricity uniformly distributed throughout a sphere" (p. 670). Rutherford explicitly points out that Crowther's experimental results provided support for Thomson's hypothesis of compound scattering:

> The theory of Sir J. J. Thomson is based on the assumption that the scattering due to a single atomic encounter is small, and the particular structure assumed for the atom does not admit of a very large deflexion of an alpha particle in traversing a single atom, unless it be supposed that the diameter of the sphere of positive electricity is minute compared with the diameter of the influence of the atom. (p. 670)

In the last part of the sentence Rutherford was preparing the ground for his own model of the atom.

2. Rutherford now presents the other side of the story:

> The observations, however, of Geiger and Marsden [1909] on the scattering of alpha rays indicate that some of the alpha particles must suffer a deflexion of more than a right angle at a single encounter. They found, for example, that a small fraction of the incident alpha particles, about 1 in 20,000 were turned through an average angle of 90° in passing through a layer of gold-foil about.00004 cm. thick. ... A simple calculation based on the theory of probability shows that the chance of an alpha particle being deflected through 90° is vanishingly small. In addition, it will be seen later that the distribution of the alpha particles for various angles of large deflexion does not follow the probability law to be expected if such large deflexions are made up of a large number of small deviations. It seems reasonable to suppose that the deflexion through a large angle is due to a single atomic encounter, for the chance of a second encounter of a kind to produce a large deflexion must in most cases be exceedingly small. A simple calculation shows that *the atom must be a seat of an intense electric field in order to produce such a large deflexion at a single encounter.* (p. 669, emphasis added)

This summarized, on the very first page of the article, the experimental work of Rutherford's colleagues, his hypothesis of single scattering, and a glimpse of his nuclear atom.

3. Later in the article Rutherford provides calculations and emphasizes that the probability of a second deflection is negligible:

> [I]t is assumed that the alpha particles scattered through a large angle suffer only one large deflexion. If, for example, the probability of a single deflexion Φ in passing through a thickness t is 1/1000, the probability of two successive deflexions each of value Φ is $1/10^6$, and is negligibly small. (p. 675)

4. Besides Thomson's model of the atom, Rutherford also mentions another rival, Nagaoka's (1904) "Saturnian" model of the atom, which consisted of a central attracting mass surrounded by rings of electrons. Nagaoka showed that such a system was stable if the attractive force was large. With regard to Nagaoka's model, Rutherford explained: "the chance of large deflexion would practically be unaltered, whether the atom is considered to be a disk or a sphere" (p. 688).
5. It is of interest to note that until April 1911, when this article was written, Rutherford makes no mention of:

   (a) An analogy of his model to the solar system
   (b) Nucleus, to represent the central charge of the atom
   (c) Whether the central charge was positive or negative

With respect to the central charge, Rutherford explicitly stated that "it has not so far been found possible to obtain definite evidence to determine whether it be positive or negative" (p. 688). This shows the tentative nature of Rutherford's model of the atom.

Heuristic Principles of Rutherford's Model of the Nuclear Atom

In this section we consider information available in the literature that could facilitate a reconstruction of the events leading to the postulation of Rutherford's model. Apparently, Rutherford had the experimental data as early as June 1909 (Geiger & Marsden, 1909) to postulate his model of the nuclear atom. It is of interest to reconstruct the events that finally led Rutherford to announce his model on March 7, 1911, at a meeting of Manchester Literary and Philosophical Society, and ultimately to publish them in the *Philosophical Magazine* in May 1911. What happened between June 1909 and March 1911? In his presidential address to the annual meeting of the British Association, Winnipeg, Canada, held in the summer of 1909, Rutherford was referring to the recent article of Geiger and Marsden (1909), when he reported:

> The conclusion is unavoidable that the atom is the seat of an intense electric field, for otherwise it would be impossible to change the direction of the particle in passing over such a minute distance as the diameter of a molecule. (Reproduced in Wilson, 1983, p. 292)

Crowther (1910) published experimental findings that provided evidence for Thomson's (1910) hypothesis of compound scattering of alpha particles. This apparently forced Rutherford, Geiger, and Marsden to do further experiments before facing the challenge of Thomson and colleagues. According to Wilson (1983):

> J. J. [Thomson] had people working on the scattering problem in his own laboratory, and a paper by one of his men, Crowther [1910], became of crucial importance in the battle between the two concepts of the atom. It is, however, too often ignored that Rutherford's superior concept of atomic structure also involved the overthrow of his master's [Thomson] model. (p. 295)

In a series of letters written to friends and colleagues, just before announcing his hypothesis of the nuclear atom on March 7, 1911, Rutherford acknowledges the serious challenge posed by the rival hypothesis; namely, Thomson's hypothesis of compound scattering, based on Crowther's experimental work. Following are excerpts of Rutherford's letters:

> Dec. 14, 1910: I think I can devise an atom much superior to J. J.'s [Thomson] for the explanation of and stoppage of alpha- and beta-particles, and at the same time I think it will fit in extraordinarily well with the experimental numbers. It will account for the reflected alpha-particles observed by Geiger and generally I think will make *a fine working hypothesis*. (Letter to Boltwood, reproduced in Wilson [1983], p. 295, emphasis added)

> Feb. 8, 1911: [Geiger's results] look very promising for the theory. I am beginning to think that the central core is negatively charged, for otherwise the law of absorption for beta-rays would be very different from that observed. (Letter to Bragg, reproduced in Wilson [1983], p. 300)

> Feb. 9, 1911: I have looked into Crowther's scattering paper carefully, and the more I examine it the more I marvel at the way he made it fit (or thought he made it fit) J. J.'s theory. ... Altogether I think the outlook is decidedly promising. (Letter to Bragg, reproduced in Wilson [1983], p. 300)

> Feb. 11, 1911: I am quite sure the numbers of the earlier part of the curve [Crowther's] were fudged. (Letter to Bragg, reproduced in Wilson [1983], pp. 300–301)

> March 8, 1911: I may mention that the theory of large [single] scattering will hold equally well if instead of one large central charge one supposed the atom to consist of a very large number of smaller charges distributed throughout the atom. It can be shown however that, on this view, the small scattering should be much greater than that experimentally observed. It is consequently simplest to consider the effect of a single point charge. (Letter to Madsen, reproduced in Wilson [1983], p. 302)

The last letter, written the day after his address to the Manchester Literary and Philosophical Society on March 7, 1911, is indeed a strange demonstration of ambivalence. In a sense it shows the power and acceptance of the rival theory (Thomson's compound scattering) in the scientific community. On the one hand, Rutherford was entirely convinced and optimistic that his model of the atom explained experimental findings better, yet on the other hand it seems that the prestige, authority, and even perhaps some reverence for his teacher made him waver. However, in a letter to Schuster (Secretary of the Royal Society), written about 3 years later (February 2, 1914), Rutherford is much more forceful:

> I have promulgated views on which J. J. [Thomson] is, or pretends to be, sceptical. At the same time I think that if he had not put forward a theoretical atom himself, he would have come round long ago, for the evidence is very strongly against him. If he has a proper scientific spirit I do not see why he should hold aloof and the mere fact that he was in opposition would liven up the meeting. (Reproduced in Wilson [1983], p. 338)

Interestingly, the scientific community received Rutherford's model with some caution. No notice appeared in *Nature, Physikalische Zeitschrift,* or *Rapports* of the Solvay Congress of 1911, which was attended by Rutherford himself. Thomson himself understandably did not mention it in the series of lectures he delivered on the "Structure of the Atom" at the Royal Institution in 1913. For most contemporary physicists, based on the evidence available, it was difficult to take a decision in favor of Thomson's hypothesis of compound scattering or Rutherford's hypothesis of single scattering. This clearly shows that data from experiments do not unambiguously lead to the formulation of theories, and similar experimental data can be interpreted in various different ways. Even Crowther (who provided the crucial experimental data to support Thomson's hypothesis) was uncertain as late as 1914 and suggested that the models of Thomson and Rutherford were perhaps special cases of the most probable hypothesis (Crowther, 1914). Curie (1914) also took a similar position. Crowther (1974) himself, recounting the events many years later, set the record straight:

> J. J. [Thomson] used Crowther's results in elaborating his theory of the atom as a region of positive electrification, in which electrons were distributed like plums in a pudding. Rutherford closely analysed Crowther's experiments, and concluded that they did not provide valid evidence for this model. (p. 151)

In retrospect, another aspect of Rutherford's experiments that deserves more attention is that only a very small fraction (1 in 20,000) of the alpha particles deflected through large angles. Furthermore, based on the theory of probability, Rutherford showed that: (a) the chance of an alpha particle being deflected through large angles was "vanishingly small"; and (b) the probability of an alpha particle experiencing a second deflection was "negligibly small." It was precisely for these reasons that he and others found the hypothesis of single scattering, and a model of the atom with an "intense electric field," so convincing. Interestingly, early in 1909, Rutherford enrolled to attend elementary lectures on probability given by Horace Lamb, and his notebooks bear witness to his attendance and having taken extensive notes (Wilson, 1983, p. 290). This information is important as it was generally considered that Thomson (rather than Rutherford) was more adept at mathematical modeling. Herron (1977), for example, has emphasized that students get the impression that the surprising part of these experiments was that most alpha particles passed through the gold foil undeflected, whereas

> [i]t was the 1 in 20,000 particles deflected through large angles that led Rutherford to postulate that the positive charge in the atom is concentrated in a small region of space at its center and the idea of a nuclear atom became established as the accepted theory. (p. 499)

Looking back on Rutherford's work many years later, Millikan (1947) emphasized a similar point:

> These sharp deflections, which occasionally amount to as much as 150° to 180°, lend the strongest of support to the view that the atom consists of a heavy positively charged nucleus about which are grouped enough electrons to render the whole atom neutral. But the fact that in these experiments the *alpha particle goes through 130,000 atoms without approaching near enough to this central nucleus to suffer appreciable deflection more than two or three times constitutes the most convincing evidence ... [of] ... this central nucleus*. (p. 193, emphasis added)

Finally, according to Wilson (1983):

> Rutherford's nuclear atom did not prevail because of direct evidence in its favour – it prevailed because of its extraordinarily successful explanatory power ... explanations for large areas of problems in chemistry, particularly regarding the nature of the elements and the regularities and differences between them. (p. 306)

Cooper (1992) has referred to the role of experiments and conjectures in Rutherford's efforts to construct his model of the atom: "The situation in 1910, as pieced together from these various *experiments and conjectures*, seemed to be somewhat as follows: atoms, electrically neutral, consisted of electrons (of very small mass) and positive material with the major part of the mass" (Cooper, 1992, p. 316, emphasis added; also Cooper, 1970, p. 316).

Educational Implications of the Rutherford Model of the Atom

Blanco and Niaz (1998) asked freshman students enrolled in a general chemistry course: "How would you have interpreted, if most of the alpha particles would have deflected through large angles?" The basic idea behind this item was to provide students with hypothetical experimental evidence (quite different from what they had studied), and observe the degree to which students can assimilate conflicting evidence and go beyond by formulating alternative models. Most students did not understand that as the evidence changes, models will also change and following are some of the examples of students' responses: (a) assuming that a repulsive force would impede penetration of the atom; (b) due to the fact that a majority of the alpha particles were heavy; (c) magnitude of the positive charge density would be much greater; (d) as if the nucleus of the gold atom would not be in the center but on the surface; (e) they would be deflected because on colliding with the nucleus all of them would not be able to enter; (f) as if the nucleus had a greater mass and the number of electrons much greater; (g) size of the nucleus would be relatively big and hence the electrons would have greater mass and positive charge. It is interesting to note that these students do make an effort to provide an explanation of the hypothetical experimental finding. These explanations are, however, based entirely on some other experimental property not taken into consideration previously, such as repulsive forces, greater positive charge density, and bigger size of the nucleus. There is, however, no attempt to formulate an alternative model.

Results obtained in this item show the need for discussing various aspects of an experiment including hypothetical findings so that students can discuss and debate and thus familiarize themselves with the dynamics of conflicts and controversies, and their significance in scientific practice.

In a follow-up study (Niaz et al., 2002), freshman students responded to the research question previously presented in this section, after having participated in various discussions based on a historical reconstruction of Rutherford's experiments (cf. Niaz, 1998). It is important to note that in this question students were responding to a hypothetical situation in which "most of the alpha particles would have deflected through large angles." Experimental group students not only performed better than the control group, but their responses indicated a better grasp of the new hypothetical experimental finding. These responses showed clearly that in the light of the new and contradictory evidence presented, experimental group students were willing to give up the model presented in textbooks and discussed in class. Furthermore, some of them even went beyond and proposed an alternative model of the atom, based on a hypothetical set of data and thus facilitated conceptual understanding.

To maintain his model of the atom and to explain large angle deflections of alpha particles, Thomson put forward the hypothesis of compound scattering, whereas Rutherford had explained the same experimental finding through his hypothesis of single scattering. This was the major controversy facing atomic models at the beginning of the twentieth century, and the contending parties were led by prominent scientists. Let us see how freshman science textbooks deal with such an interesting and crucial episode in the history of science. Niaz (1998) has shown that of the 23 general chemistry textbooks analyzed none described satisfactorily or mentioned this principle. Rodríguez and Niaz (2004a) have reported that of the 41 general physics textbooks analyzed, two described it satisfactorily and two made a simple mention. All textbooks analyzed were published in the USA. Once again Cooper (1970, p. 321), based on original sources, presented a detailed satisfactory presentation by arguing that Thomson's hypothesis of compound scattering was improbable and had only one chance in 10^{3500} of occurring, and hence was not accepted by the scientific community.

Bohr's Model of the Atom

Little did Bohr know what was awaiting him when he arrived in Cambridge in September 1911 to work with J.J. Thomson, then considered to be a "pioneer of the electron theory of metals and the acknowledged world master in the design of atomic models" (Heilbron & Kuhn, 1969, p. 223). Earlier in the year, in May 1911, he had successfully defended in Copenhagen his doctoral dissertation based on the electron theory of metals. Bohr had brought a rough English translation of his dissertation, which he wanted Thomson to read. After months of waiting, with the dissertation among the pile of manuscripts on Thomson's desk, Bohr decided to go and work with Rutherford in Manchester. Thus, "Thomson had the unfortunate distinction of losing for the Cavendish [Laboratory] both Rutherford and Bohr, founders of modern physics" (Snow, 1981, p. 52). Bohr arrived in Manchester in March 1912 and, after some experimental work on radioactivity, started working to quantize Rutherford's atom. In July 1912, Bohr submitted a preliminary draft to Rutherford, which he himself labeled as: "First draft of the considerations contained in the paper 'On the constitution of atoms and molecules' (written up to show these considerations to Prof. Rutherford)/(June and July 1912)." Heilbron and Kuhn (1969) consider this first draft to be a crucial document in the history of quantum theory, and refer to it as the "Rutherford Memorandum" (p. 244). Bohr's style of work was indeed unique if not enigmatic, which has led some philosophers of science to ask:

> What suddenly turned his attention from electron theory to atom models during June 1912? Why did he then choose to develop the new, little-known Rutherford atom rather than, say, the older, more successful model proposed by J. J. Thomson? Why did he approach the quantization problem in the particular way he did, one which bore impressive fruits at once and which, a year later, began to revolutionize physics? (Heilbron & Kuhn, 1969, p. 212)

Nevertheless, it took several months before Bohr submitted the final draft to Rutherford on March 6, 1913. It was accepted by the *Philosophical Magazine* on April 5 and published in July 1913 as the first part of a trilogy (Bohr, 1913a).

Bohr's (1913a) Article in the Philosophical Magazine

1. Bohr starts the first part of his trilogy with the following words: "In order to explain the results of experiments on scattering of alpha rays by matter Prof. Rutherford has given a theory of the structure of atoms" (p. 1). Next, Bohr briefly describes Rutherford's model of the atom.
2. In the next paragraph Bohr points out difficulties with Rutherford's model of the atom:

> In an attempt to explain some of the properties of matter on the basis of this atom-model we meet, however, with difficulties of a serious nature arising from the apparent instability of the system of electrons: difficulties purposely avoided in atom-models previously considered, for instance, in the one proposed by Sir J. J. Thomson. According to the theory of the latter the atom consists of a sphere of uniform positive electrification, inside which the electrons move in circular orbits. (p. 2)

3. In the third paragraph Bohr makes a comparison of the Thomson and Rutherford models:

> The principal difference … consists in the circumstance that the forces acting on the electrons in the atom-model of Thomson allow of certain configurations and motions of the electrons for which the system is in a stable equilibrium; such configurations, however, apparently do not exist for the second atom-model. (p. 2)

It is important to observe that, although the Thomson model could not be sustained after Rutherford's alpha particle experiments, it did provide for stability.

4. In the fourth paragraph Bohr formulates his epoch-making postulate:

> The way of considering a problem of this kind has, however, undergone essential alterations in recent years owing to ... experiments on very different phenomena such as specific heats, photoelectric effect, Röntgen-rays, etc. The result of the discussion of these questions seems to be a general acknowledgement of the inadequacy of the classical electrodynamics in describing the behaviour of systems of atomic size ... it seems necessary to introduce in the laws in question a quantity foreign to the classical electrodynamics, i.e., Planck's constant, or as it often is called the elementary *quantum of action*. (p. 2, emphasis added)

5. Next, Bohr points out that based on his model, it is possible to take into account the law of the line spectrum of hydrogen, leading to the Balmer, Paschen, and the other series. Although the details of Bohr's calculations for energy emission during electron transition are well known, this aspect of his theory has been the subject of considerable research and controversy in the philosophy of science literature, and will be discussed in the next section.
6. Bohr constantly refers to a rival theory by J.W. Nicholson (1911, 1912), which also presented a quantized version of the atomic model:

> Nicholson has obtained a relation to Planck's theory showing that the ratios between the wavelength of different sets of lines of the coronal spectrum can be accounted for with great accuracy by assuming that the ratio between the energy of the system and the frequency of rotation of the ring is equal to an entire multiple of Planck's constant. (p. 6)

Bohr then presents a considerably detailed critique of Nicholson's model. The important point is that Bohr had to contend with a rival model. Details of this challenge to Bohr's model will be discussed in the next section.

Heuristic Principles of Bohr's Model of the Atom

Interpretation of Bohr's model of the atom by philosophers of science is very instructive. Bohr's main objective was to explain the paradoxical stability of the Rutherford atom, and still most textbooks consider Bohr's major contribution to be the explanation of the Balmer and Paschen series of the hydrogen line spectrum. According to Lakatos (1970):

> Bohr's problem was not to explain Balmer's and Paschen's series, but to explain the paradoxical stability of the Rutherford atom. Moreover, Bohr had not even heard of these formulae before he wrote the first version of his paper. (p. 147)

The "first version" Lakatos is referring to is of course the "Rutherford Memorandum" written in June–July 1912 (cf. Heilbron & Kuhn, 1969, p. 244). A letter written by Bohr to Rutherford on January 31, 1913, shows that even then he was not fully aware of the implications of spectroscopic research for his problem:

> I do not at all deal with the question of calculation of the frequencies corresponding to the lines in the visible spectrum. I have only tried, on the basis of the simple hypothesis, which

> I used from the beginning, to discuss the constitution of the atoms and molecules in their "permanent state." (Reproduced in Rosenfeld [1963], pp. xxxvi–xxxvii)

Actually, Bohr was quite skeptical about the relevance of spectra for his model of the atom. Many years later, in an interview with Thomas Kuhn in 1962, Bohr expressed this quite explicitly:

> The spectra was a very difficult problem. ... Just as if you have the wing of a butterfly, then certainly it is very regular with the colors and so on, but nobody thought that one could get the basis of biology from the coloring of the wing of a butterfly. (Reproduced in Heilbron & Kuhn, 1969, p. 257)

Apparently, it was the spectroscopist, H.M. Hansen who familiarized Bohr with the spectroscopic work and its implications for his model (cf. Jammer, 1966, p. 77). Having seen the importance of the spectra Bohr is said to have repeated often: "As soon as I saw Balmer's formula, the whole thing was immediately clear to me" (reproduced in Rosenfeld [1963], p. xxxix). Interestingly, Kuhn points out that even before the "Rutherford Memorandum" was discovered in Bohr's files, he had conjectured

> that Bohr had developed a detailed, non-spectroscopic, quantized version of Rutherford's atom some time before he saw the relevance of the Balmer formula. (Heilbron & Kuhn, 1969, p. 255)

A reconstruction of these events related to the development of the structure of the atom provides an understanding how science progresses and is practiced. Lakatos (1970) has shown the importance of these events in the history of science in the following terms:

> Since the Balmer and the Paschen series were known before 1913 [year of Bohr's first publication], some historians present the story as an example of a Baconian "inductive ascent": (1) the chaos of spectrum lines, (2) an "empirical law" (Balmer), (3) the theoretical explanation (Bohr). (p. 147)

Lakatos clearly uses this episode in the history of science to emphasize that science does not proceed from experimental observations to scientific laws and theories, through inductive generalizations. In spite of their many differences, most new philosophers of science would agree to this conceptualization of scientific progress (cf. Feyerabend, 1970; Hanson, 1958; Kuhn, 1970; Lakatos, 1970; Laudan, 1977). This perspective based on the new philosophy of science can be summarized as:

> The role of observation is not to provide a firm basis from which generalizations can then be inductively extrapolated but, if anything, to provide some check on whether the promise of previously made theoretical commitments has been fulfilled. (Papineau, 1979, p. 100)

Once again, Lakatos (1970) expresses the argument cogently:

> [T]he progress of science would hardly have been delayed had we lacked the laudible trials and errors of the ingenious Swiss school-teacher [Balmer]: the speculative mainline of science, carried forward by bold speculations of Planck, Rutherford, Einstein and Bohr would have produced Balmer's results deductively, as test-statements of their theories, without Balmer's so-called "pioneering". In the rational reconstruction of science there is little reward for the pains of the discoverers of "naive conjectures". (p. 147)

An important aspect of Bohr's model of the atom is the presence of a deep philosophical chasm: that is, in the stationary states, the atom obeys classical laws of Newtonian mechanics; on the other hand, when the atom emits radiation, it exhibits discontinuous (quantum) behavior, according to laws first proposed by Planck in 1900. Rutherford, although no philosopher of science, was the first to point this out, when he wrote to Bohr on March 20, 1913:

> [T]he mixture of Planck's ideas with the old mechanics makes it very difficult to form a physical idea of what is the basis of it all. ... How does the electron decide what frequency it is going to vibrate at when it passes from one stationary state to another? (Reproduced in Holton, 1993, p. 80)

Bohr's 1913 article, in general, had a fairly adverse reception in the scientific community. Otto Stern told a friend: "If that nonsense is correct which Bohr has just published, then I will give up being a physicist" (reproduced in Holton, 1986, p. 145). Lord Rayleigh was categorical: "It does not suit me" (reproduced in Holton, 1993, p. 79). H.A. Lorentz objected: "[T]he individual existence of quanta in the aether is impossible" (reproduced in Holton, 1993, p. 79). J.J. Thomson, whom Bohr considered as the "world master in the design of atomic models" objected to Bohr's conception in most of his writings from 1913 to 1936 (cf. Holton, 1993, p. 79).

More recently, philosophers of science have been more understanding of Bohr's model of the atom:

> [I]t is understandable that, in the excitement over its success, men overlooked a malformation in the theory's architecture; for Bohr's atom sat like a baroque tower upon the Gothic base of classical electrodynamics. (Margenau, 1950, p. 311)
>
> Thus Bohr's atom of 1913 was really a kind of mermaid – the improbable grafting together of disparate parts, rather than a new creation incorporating quantum theory at its core. (Holton, 1986, p. 145)

These appreciations (Margenau and Holton), written many years after Bohr's publication in 1913, still lack a historical perspective, as to what exactly Bohr was doing. Lakatos, on the other hand, shows that Bohr used a methodology frequently employed by scientists in the past and perfectly valid for the advancement of science:

> *[S]ome of the most important research programmes in the history of science were grafted on to older programmes with which they were blatantly inconsistent.* For instance, Copernican astronomy was "grafted" on to Aristotelian physics, Bohr's programme on to Maxwell's. Such "grafts" are irrational for the justificationist and for the naive falsificationist, neither of whom can countenance growth on inconsistent foundations. ... As the young grafted programme strengthens, the peaceful co-existence comes to an end, the symbiosis becomes competitive and the champions of the new programme try to replace the old programme altogether. (Lakatos, 1970, p. 142, italics in original)

At this stage it would be interesting to observe how Leon Cooper (Nobel Laureate, physics, 1972) has described Bohr's contribution:

> In 1913 Niels Bohr proposed his famous theory of the hydrogen atom. One cannot say that he resolved the problems raised by Rutherford. In a sense he crystallized the dilemma in an even more dramatic form. Focusing his attention entirely on the construction of a nuclear atom, Bohr took what principles of classical physics he needed and added several

> nonclassical hypotheses almost without precedent; the mélange was not consistent. But they formed a remarkably successful theory of the hydrogen atom. It would be years before it could be said that one had a consistent theory again. (p. 325, Cooper, 1970)

Cooper is not a philosopher of science and still recognized that although the mélange was not consistent, it was still a successful theory. More recently, Cooper has elaborated on the role of inconsistencies in Bohr's theory: "But the inconsistencies might lead to new theoretical ideas. This is exactly what happened in the case of the Bohr atom; there were obvious inconsistencies, but the Bohr atom was so successful that the fruitful approach was to find out how to remove the inconsistencies and still retain the structure of the theory" (reproduced in Niaz et al., 2009).

Educational Implications of the Bohr Model of the Atom

Blanco and Niaz (1998) asked freshman students enrolled in a general chemistry course, the following question: "If Bohr changed Rutherford's model of the atom, did Rutherford make mistakes while doing his experiments?" The idea behind this item was to make students think and reflect with respect to progress in science, namely we do not necessarily replace wrong theories with right ones, but rather look for greater explanatory power. Following examples (from students' responses) show that most of the students did not understand this important aspect of the dynamics of scientific progress: (a) no, what happened was that Rutherford only limited himself to the existence of the nucleus; (b) no, he only made mistakes with respect to the orbitals of each electron; (c) no, in reality he made deductions from what he observed.; (d) yes, because he established that the electrons circled around the nucleus and according to the laws of classical physics the electron would have lost energy, and ultimately destroyed the atom; (e) yes, some phenomenon did not permit him to observe the *real model of the atom* (emphasis added); (f) yes, at least he made some mistakes, otherwise Bohr would not have taken the care to present a good model; (g) yes, because he *based his model on suppositions* (emphasis added). Some of the students provided a better understanding and following are some examples: (a) no, Rutherford's experiments served as a base for Bohr to carry out his experiments; (b) no, each experiment contributes towards the perfection of a previously obtained result; (c) no, each one of them made his postulations according to their knowledge and the progress in the field at that time. Some of the responses to this item seem to emphasize that Rutherford based his model on suppositions and not hard facts, and hence the need for Bohr's model. This makes interesting reading, as Bohr's four postulates were based not only on suppositions (presuppositions, according to Holton, 1978) but also on speculations, which according to Lakatos is perfectly justifiable, and helped Bohr to solve the paradox of the stability of the Rutherford atom by postulating a rival research program.

In a follow-up study (Niaz et al., 2002) freshman students were asked the same question as previously presented in this section. These students participated in classroom discussions based on a historical reconstruction of atomic models (Niaz, 1998).

Experimental group students performed much better than the control group and the following responses provide evidence for their greater understanding of how scientific theories develop: (a) on the contrary, Rutherford's experiments constituted a base, which was later used by Bohr to establish his model … scientific process is characterized by the perfection of established theories … thus Rutherford's experiments were the points of departure for new discoveries; and (b) the model proposed by Rutherford constituted an extraordinary scientific advance for that time … but just as all models are subject to further tests by different scientists, Rutherford's model was changed by Bohr's. It is plausible to suggest that these responses come quite close to what philosophers of science would refer to as the tentative nature of science (Lakatos, 1970).

Bohr's incorporation of Planck's "quantum of action" into the classical electrodynamics of Maxwell represented a strange "mixture" for many of Bohr's contemporaries and philosophers of science. Lakatos (1970) has explained how when faced with difficulties scientists (Bohr's example is cited explicitly) end up constructing a new research program based on the existing one. This illustrates how scientists resort to contradictory "grafts." This is a good opportunity to illustrate another facet of the dynamics of progress in science, and it would be interesting to observe how textbook authors handle such episodes in the history of science. Niaz (998) has reported that of the 23 general chemistry textbooks analyzed, four simply mentioned this heuristic principle and two made a satisfactory presentation of which the following is an example:

> There are two ways of proposing a new theory in science, and Bohr's work illustrates the less obvious one. One way is to amass such an amount of data that the new theory becomes obvious and self-evident to any observer. The theory then is almost a summary of the data. The other way is to make a bold new assertion that initially does not seem to follow from the data, and then to demonstrate that the consequences of this assertion, when worked out, explain many observations. With this method, a theorist says, "You may not see why, yet, but please suspend judgment on my hypothesis until I show you what I can do with it." Bohr's theory is of this type. Bohr said to classical physicists: "You have been misled by your physics to expect that the electron would radiate energy and spiral into the nucleus. Let us assume that it does not, and see if we can account for more observations than by assuming that it does." (Dickerson et al., 1984, p. 264)

Among other positive features, it is important to recognize that these authors provide students with alternatives with respect to theory construction.

Rodríguez and Niaz (2004a) have reported that of the 41 general physics textbooks analyzed, only one described satisfactorily that Bohr's incorporation of Planck's "quantum of action" was based on an inconsistent foundation and represented a "deep philosophical chasm." Once again it was Cooper (1970), who provided an example of a satisfactory presentation. At this stage, it would be appropriate to pause and reflect why textbook authors ignore the historical record and provide students a vision of the "science in the making" that comes quite close to naivety.

Justi and Gilbert (2000) analyzed high school chemistry textbooks (nine from Brazil and three from the UK) to study the presentation of atomic models. These authors report the use of hybrid models based on various historical developments, such as Ancient Greek, Dalton, Thomson, Rutherford, Bohr, and Quantum mechanics

(Schrödinger's equation). Based on their findings these authors concluded that hybrid models, by their very nature as composites drawn from several distinct historical models, do not allow the history and philosophy of science to make a full contribution to science education.

Leary and Kippeny (1999) have presented a historical outline (Faraday to Chadwick) of the development of the modern atom and suggested that students must be challenged to think critically, if eventually they are to be entrusted with the responsibility for societal and scientific advancement.

Chapter 7
Determination of the Elementary Electrical Charge: Millikan and Ehrenhaft*

Introduction

Most chemistry and physics textbooks consider the oil drop experiment to be a simple, classic, and beautiful experiment, in which Robert A. Millikan (1868–1953) by an exact experimental technique determined the elementary electrical charge. Polanyi (1964) has emphasized the degree to which established knowledge in textbooks departs from the events associated with the original discovery:

> Yet as we pursue scientific discoveries through their consecutive publication on their way to the textbooks, which eventually assures their reception as part of established knowledge by successive generations of students, and through these by the general public, we observe that the intellectual passions aroused by them appear gradually toned down to a faint echo of their discoverer's first excitement at the moment of Illumination. … A transition takes place here from a heuristic act to the routine teaching and learning of its results, and eventually to the mere holding of these as known and true, in the course of which the personal participation of the knower is altogether transformed. (pp. 171–172)

Analyses of chemistry and physics textbooks shows that Polanyi (1964) had indeed foreseen the dilemma with much acumen (Matthews, 1994; Niaz, 2000b; Rodríguez & Niaz, 2004c). A historical reconstruction of the events that led to the determination of the elementary electrical charge (*e*) shows the controversial nature of the oil drop experiment then (1910–1925) and that the experiment is difficult to perform even today (Jones, 1995). The experiment itself has been accepted enthusiastically in many circles without much critical scrutiny. In a poll conducted for *Physics World*, its readers considered the oil drop experiment to be one of the ten 'most beautiful' ones of all time (Crease, 2002). Furthermore, according to Crease many respondents considered that the experiment was conceived, carried out, and understood with considerable ease. Similarly, Segerstrale (1995) contrasted the opposing interpretations of Millikan's research

* Reproduced, in part from Niaz, M. (2005b). An appraisal of the controversial nature of the oil drop experiment: Is closure possible? *British Journal for the Philosophy of Science, 56,* 681–702.

M. Niaz, *Critical Appraisal of Physical Science as a Human Enterprise: Dynamics of Scientific Progress*, Science and Technology Education Library 36,

ethics, which show the tendency to produce "canned" stories about Millikan that are stereotyped and oversimplified.

Acceptance of the quantization of the elementary electrical charge was preceded by a bitter dispute between Millikan (University of Chicago) and Felix Ehrenhaft (University of Vienna, 1879–1952) that lasted for many years (1910–1925). Both Millikan and Ehrenhaft obtained very similar experimental results and yet Millikan was led to formulate the elementary electrical charge (electron) and Ehrenhaft to fractional charges (sub-electron). Interestingly, however, after almost 90 years historians and philosophers of science do not seem to agree as to what really happened. A review of the literature shows that there have been at least five major attempts to reconstruct the historical details that led to the Millikan–Ehrenhaft controversy:

1. Holton (1978)
2. Franklin (1981)
3. Barnes, Bloor, and Henry et al. (1996)
4. Goodstein (2001)
5. Hintikka (2005)

Interpretations of Holton (1978), Franklin (1981), and Goodstein (2001) had the additional vantage point of having consulted Millikan's handwritten notebooks available at the Millikan Archives, California Institute of Technology, Pasadena. This shows that with respect to the oil drop experiment we have the controversy not only among the original protagonists but also among those who have tried to understand and interpret the experiment. According to Machamer, Pera and Baltas (2000), although most of the achievements in scientific progress have involved controversies, it is paradoxical that there is dissociation between science as actually practiced and science as perceived by both scientists and philosophers:

> While nobody would deny that science in the making has been replete with controversies, the same people often depict its essence or end product as free from disputes, as the uncontroversial rational human endeavor par excellence. (p. 3)

The objective of this chapter is a critical appraisal of the five interpretations with respect to the controversial nature of the oil drop experiment and an attempt to provide a closure to the controversy. This historical reconstruction is presented in the following sections:

1. An appraisal of Holton's interpretation
2. An appraisal of Franklin's interpretation
3. An appraisal of Barnes, Bloor, and Henry's interpretation
4. An appraisal of Goodstein's interpretation
5. A crucial test: The second drop (reading) of 15 March, 1912
6. An appraisal of Hintikka's interpretation
7. Educational implications of the historical reconstruction of the oil drop experiment
8. Conclusion: Is closure possible?

An Appraisal of Holton's Interpretation

One of the most important aspects of Holton's (1978) interpretation is that he attributes Millikan's data selection and reduction procedures to his presuppositions with respect to the atomic nature of electricity. The importance of presuppositions in a research program has also been recognized by others (e.g., "hard-core" Lakatos, 1970; "guiding assumptions" Laudan et al., 1988; see Chapter 1 for more details). Holton highlights the importance of presuppositions by drawing attention to how Millikan in his first major publication (Millikan, 1910, manuscript submitted 9 October, 1909) started with the following words:

> Among all physical constants there are two which will be universally admitted to be of predominant importance; the one is the velocity of light, ... and the other is the ultimate, or elementary, electrical charge, a knowledge of which makes possible a determination of the absolute values of all atomic and molecular weights, the absolute number of molecules in a given weight of any substance, the kinetic energy of agitation of any molecule at a given temperature, and a considerable number of other important physical quantities. (Millikan, 1910, p. 209)

This shows that Millikan had a fairly good idea as to what he was looking for, even before the controversy with Ehrenhaft started in 1910. In later publications, Millikan (1916) explicitly outlined "The History of the Idea of a Unit Charge." Besides Benjamin Franklin, the origin of the idea is attributed to Faraday's discoveries in electrolysis (1833), Stoney's use of the word *electron* in 1874, Thomson's (1897) work at the Cavendish Laboratory, and finally Millikan concluded:

> They did ... give great stimulus to the atomic theory of electricity and caused it to become the prevalent mode of interpreting electrical phenomena. They brought to light the existence of a body, J.J. Thomson's corpuscle, for which the value of *e/m* was 1/1830 of that found on the hydrogen ion in electrolysis. Townsend [1897], J.J. Thomson [1898], H.A. Wilson [1903], Przibram [1907], Millikan and Begeman [1908], Ehrenhaft [1909] and [De] Broglie [1909] in succession made rough determinations or estimates of the average charge appearing on gaseous ions and found it equal, within the limits of uncertainty ... to the value estimated for the univalent ions in electrolysis. (Millikan, 1916, p. 596)

Inclusion of Ehrenhaft and his colleague Przibram in this list of researchers supporting the atomic theory of electricity may sound strange. Holton (1978) shows how in his earlier work Ehrenhaft (1909) did subscribe to what he referred to as the "Elementary Quantum of Electricity" (p. 185).

The second important aspect of Holton's (1978) interpretation is that Millikan was not only convinced about the atomic theory of electricity but also about the magnitude of the elementary electrical charge. In Holton's opinion, given the pioneering work of Thomson and colleagues, there was enough evidence to be convinced "... of the particle theory of unitary electrical charge, even prior to Millikan's work" (p. 180). Millikan could foresee the next logical step – determination of the unitary charge. Millikan and Begeman (1908) based on work done in 1907 using the cloud method reported a value of 4.03×10^{-10} esu for the unitary charge. This value was quoted by Rutherford and Geiger (1908) and they considered it to be one of the best experimental values available. Rutherford and Geiger had determined the magnitude of the alpha particle charge as 9.3×10^{-10} esu and assumed that it should be equal to $|2e|$ and hence e (unitary charge) should be 4.65×10^{-10} esu. According to Millikan and Begeman (1908) the value of 4.03×10^{-10} esu was 15% lower, and from then on the value suggested by Rutherford and Geiger was a sort of guidepost for Millikan and many other researchers.

The third important aspect of Holton's (1978) interpretation is to have studied Millikan's handwritten notebooks at CALTECH, Pasadena. The notebooks have data from the period 28 October, 1911, to 16 April, 1912, consisting of about 175 pages. Data from these notebooks were used to publish Millikan (1913), a major publication in defense of his methodology. According to Holton's account there were 140 experiments on an equal number of oil drops. In the actual publication complete data on 58 drops is meticulously presented and it is emphasized that

> *[i]t will be seen from Figs. 2 and 3 that there is but one drop in the 58 whose departure from the line amounts to as much as 0.5 percent. It is to be remarked, too, that this is not a selected group of drops but represents all of the drops experimented upon during 60 consecutive days.* (Millikan, 1913, p. 138, original italics)

How do we interpret this information? The laboratory notebooks tell us that there were 140 drops and the published results are emphatic that there were 58 drops. What happened to the other 82 (59%) drops? Herein lies the crux of the difference in the methodologies of Ehrenhaft and Millikan. In other words, Millikan perhaps excluded drops that did not have charge equal to an integral multiple of the elementary electrical charge e, as suggested by the work of Rutherford and Geiger (1908). Holton (1978) wondered about Ehrenhaft's response if he had had access to Millikan's notebooks:

> If Ehrenhaft had obtained such data, he would probably not have neglected the second observations [that did not give the expected value of e] and many others like it in these two notebooks that shared the same fate; he would very likely have used them all. (pp. 209–210)

At this stage, it is important to note that Ehrenhaft, too, obtained data that he interpreted as integral multiple of the elementary electrical charge (e). Nevertheless, his argument was precisely that there were many drops that did not lead to an integral multiple of e. The crux of the issue is: What was the warrant under which Millikan discarded more than half of his observations? The answer is simple if we are willing to give up the so-called scientific method according to which experimental data inevitably lead to theoretical formulations. In other words, Millikan's presuppositions (existence of an elementary electrical charge and the idea that this charge must have a value within a certain range) provided the warrant. Indeed, according to Holton, Millikan would perhaps have liked to warn Ehrenhaft that all their data cannot be used as their experiments were constantly faced with difficulties such as evaporation, changes in sphericity, radius and density of droplets, validity of Stokes' law, and other experimental variables (battery voltages, stopwatch errors, temperature, pressure, convection, and so on).

Finally, Holton (1978) makes the following important observation with respect to Millikan's handling of the data: "It appears likely that after almost every run

Millikan made some rough calculations of *e* on the spot, and often he appended a summary judgment" (p. 211). As we shall see in the next section, this procedure makes the decision with respect to how and why Millikan selected some drops and not others extremely difficult. In other words, Millikan could have decided to exclude a drop as the first observations may have provided sufficient indication that it may not come close to the "expected correct" value of *e*.

An Appraisal of Franklin's Interpretation

Franklin (1981) also consulted Millikan's handwritten notebooks and his report was published in the same journal as that of Holton (1978). Some of the salient features of his interpretation are discussed in this section. Franklin reports (p. 187) that the notebooks covering the period 28 October 1911, to 16 April 1912, contain data on 175 drops. There is no explanation as to why this number differs from that (140 drops) provided by Holton (1978).

According to Franklin (1981): "Millikan had a good idea of the results he expected although his expectations seem to have changed as he went along. ... For the early events ... he was comparing his observations to the value of $e = 4.891 \times 10^{-10}$ esu given in his paper of 1911" (p. 192). Franklin ignores that besides the value of *e* from his 1911 paper, Millikan was guided by the value of *e* suggested by Rutherford and Geiger (1908). Furthermore, it is not clear what Franklin means by "his expectations seem to have changed as he went along." On the contrary, if there was one thing that could characterize Millikan's approach it was his continued belief and perseverance in his original expectations (presuppositions), viz., atomic nature of electricity and a value of *e* that approximated to that of Rutherford and Geiger (1908). Interestingly, Franklin makes no mention of this reference.

Franklin tries to convince the reader that Millikan's data before February 13, 1912, should not be considered as he was still trying to get his apparatus to work properly – perhaps a warm-up period. Based on Franklin (1981) the oil drops could be classified as follows:

(i) Total number of drops in the notebooks – 175
(ii) Number of drops studied before February 13, 1912 – 68
(iii) Number of valid drops – 107
(iv) Of the valid drops, number published by Millikan – 58
(v) Of the valid drops, number excluded by Millikan – 49
(vi) Of the valid drops number excluded with no calculation – 22
(ii) Of the valid drops number excluded after calculating *e* – 27

Reasons for excluding 27 valid drops even after having calculated the value of *e*:

- 12 drops: their values of pressure and radius were such as to require second-order correction to Stokes' law.
- 2 drops: on experimental grounds.

- 5 drops: they had three or fewer reliable measured values of t_f (represents the time of rising of the drop under the field).
- 2 drops: for no apparent reason, probably because Millikan did not need them for calculating *e*.
- 1 drop: anomalous (Franklin provides no further information)
- 5 drops: according to Franklin: "His only evident reason for rejecting these five events is that their values did not agree with his expectations" (p. 195).

Exclusion of 68 drops (studied before February 13, 1912), on the grounds that presumably the apparatus was not working well is questionable. Apparently, Millikan made no explicit qualification in the notebooks, nor did Holton allude to this aspect. Furthermore, given Millikan's general procedure of making "rough calculations of *e* on the spot" (Holton, 1978, p. 211) it is extremely doubtful that all 68 drops were excluded due to difficulties with the apparatus. It is plausible to suggest that Millikan may have excluded a drop as initial observations could have given him a rough estimate of the value of *e*.

The last five drops make an interesting case. Franklin attributes their exclusion to values that did not agree with Millikan's expectations. So in the ultimate analysis, Millikan's expectations (could be presuppositions?) were after all important. Nevertheless, Franklin does not tell us anything explicitly about Millikan's expectations. Of the last 27 drops excluded after having calculated the value of *e*, one could accept Franklin's advice with respect to the first 12 (required second-order correction to Stokes' law). For the other 15 drops there was apparently no plausible reason, except that the value of *e* was not the expected one.

Does Franklin's selective analysis of oil drops convince the reader? I am afraid not. Let us summarize all the information. There were 175 drops in the notebooks, 68 were excluded as the apparatus was presumably not working well, another 49 were excluded even when the warm-up period was over, and of these 22 were excluded for reasons that are not very clear, and of the other 27, at least 15 were excluded due to an unexpected value of *e*. The reader may now be in a greater quandary than after having read Holton (1978). It appears that Millikan excluded not 59% (82 out of 140, cf. Holton, 1978) of the drops, but rather 67% (117 out of 175, cf. Franklin, 1981) of the drops. In order to convince the reader, Franklin also plotted (Fig. 4) the value of *e* against the number of the drop studied (order of event). Franklin used the recalculated value of *e* for all the published (58) drops and unpublished (25) drops. The reader, however, will recall that there were 49 unpublished drops after February 13, 1912, and of these Millikan did not calculate the value of *e* for 22 drops – apparently for experimental reasons. One could ask why in Fig. 4, all the 49 unpublished drops were not included. This could have helped to make Millikan's (and also Franklin's) case stronger. What is even more troublesome is the fact that 13 of the 25 drops included in Fig. 4 formed part of the 22 drops for whom Millikan did not calculate the value of *e*. Franklin calculated a mean value of *e* based on his recalculation of the 58 published drops to be 4.777×10^{-10} esu (statistical error = 0.003), and for the 25 unpublished drops a mean value of 4.789×10^{-10} esu $\pm$ 0.007. Although Franklin's selection of drops helped to decrease the statistical error, one could argue that given the controversy with respect to the selection of drops by Millikan, readers would have liked to see the scatter in Fig. 4, if it had been based on the 49 unpublished drops. With respect to the statistical error in Millikan's data it is important to note Hintikka's (2006) observation: "But didn't Millikan's omission of data belie his statistical estimates of error? This is far from obvious from a methodological standpoint. Statistical methods are designed to deal with random errors, not systematic ones. Hence, if Millikan had a good reason that this or that observation was due to a systematic error in that particular observation, he would have been justified in omitting it also from his statistical error estimation."

According to Franklin (1981): "Millikan's cosmetic surgery touched 30 of the 58 published events, from which he excluded one or more (usually less than three) observations. For example, in the case of drop No. 15, Millikan used only eight of the twelve measurements of t_f in calculating *e*" (p. 197). Franklin plots in Fig. 5 Millikan's calculated values of *e* (with cosmetic surgery) and his own recalculated values for all 58 published drops and concludes that "the result of his [Millikan's] tinkering is to reduce the statistical error rather than to change the mean value of *e* (fig. 5)" (p. 198). Thus, Millikan not only excluded drops (cf. Holton, 1978) but also resorted to cosmetic surgery and selective analysis of data (cf. Franklin, 1981).

Franklin refers to three anomalous events (drops) and of these the second drop of April 16, 1912, worries him the most:

> This event is the most worrisome of the three since it is among Millikan's very best observations, as shown by internal consistency of the values. ... With a second-order correction to Stokes' law, $e = 2.810 \times 10^{-10}$ esu... both the charge on the drop, as well as the changes in charge, must be fractional, a highly unlikely occurrence. Once again neither dust nor voltage problems can explain the anomaly. Millikan remarked, 'Something wrong w[ith] therm[ometer],' but there is no temperature effect that could by any stretch of the imagination explain a discrepancy of this magnitude. Millikan may have excluded this event to avoid giving Ehrenhaft ammunition in the controversy over the quantization of charge. (Franklin, 1981, pp. 200–201)

This is indeed quite revealing and shows how a discrepant event could appear even on the last day of the notebook readings, which weakens Franklin's argument of having excluded 68 drops studied before February 13, 1912. Furthermore, this is the only explicit reference (besides a brief mention on pp. 191–192) to the controversy with Ehrenhaft. Hentschel (2004) has argued convincingly that Franklin seems to downplay the importance of such drops on the grounds that these could have come into the hands of Ehrenhaft and thus provided support for the existence of fractional charges. On the contrary, Hentschel suggests that in the debate between "rationalists"/ realists and constructivists, such episodes in the history of science must be addressed with particular care and up front rather than be tucked away in the endnotes.

One of the least convincing aspects of Franklin's (1981) study is his repeated defense of exclusion of drops by Millikan, on the grounds that he had more data than he needed for calculating *e*. He uses this argument explicitly on at least five different occasions: (1) p. 192, lines: 8–10; (2) p. 192, lines: 20–22; (3) p. 194, lines 13–15; (4) p. 194, last two lines; (5) p. 195, lines: 19–20, refers to two of the 27 drops that Millikan excluded after February 13, 1912, and Franklin states that Millikan "[excluded] two for no apparent reason, probably because he did not need them for calculating *e*." Given the controversy with Ehrenhaft it is difficult to imagine how Millikan could throw away data.

Finally, it is important to note that Franklin ignored (except for a brief mention) that not only did the Millikan–Ehrenhaft controversy play an important part in the determination of the elementary electrical charge, but that Ehrenhaft was also implicitly guided by his presuppositions (anti-atomist empiricist framework).

An Appraisal of Barnes, Bloor, and Henry's Interpretation

Barnes, Bloor, and Henry (1996) outline their plan of action by recognizing the complexity of the experiment in the following terms:

> Was there a smooth, automatic, unproblematic path joining the readings entered into the laboratory notebook and the data in the published papers that were used by Rutherford, Bohr and the rest of the scientific community? As Holton makes clear, the answer to this question is negative. The route from the notebook to the published paper was complex and interesting. (p. 22)

Next, the authors recognize that given some latitude for experimental error, charges on a series of drops can be shown to be an integral multiple of a supposed unit value, "*provided the unit is made small enough*" (p. 23, original italics). In other words, Millikan was using a prior belief (presupposition) that the unit of charge was in the region of 4.7×10^{-10} esu. At this stage, Barnes et al. (1996) ask a very pertinent question: "Were *all* of the cases in which Millikan discarded readings ones where he had grounds for suspecting the apparatus, and grounds that were independent of the readings it was producing" (p. 24, original italics). Based on the work of Holton the authors respond to this question in the negative.

With this introduction, the authors present their framework (based on Mannheim, 1952; Garfinkel, 1967) for understanding the oil drop experiment in cogent terms:

> Mannheim asked us to reflect on the predicament of someone trying to understand some fragments of a document. If they knew the import of the whole document, then they could make sense of the fragments… We can understand Millikan's work by seeing him in an analogous hermeneutic predicament, and as adopting an analogous method. His experimental data are his fragments. The whole document is the unknown reality that underlies and produces them. His guiding theory is the meaning he imputes to the document, a theory that guides his response to the evidential fragments, deciding which are reliable, which have undergone corruption or alteration or decay or misunderstanding. (Barnes et al., 1996, p. 25)

In view of the framework outlined above, the next section, entitled "A Sociological Reading," is problematic. According to the authors:

> Millikan's apparatus – as even the minimally physically informed observer can see – merely permitted electricity to interact with drops of oil. It therefore dealt with the *interface* between electricity and gross matter. To make the move from this interface to conclusions about the nature of electricity *as such* involves a special inference. What is more, this inference is one that cannot be sanctioned by the general principles of deductive reasoning nor by common sense. Not even the most unsophisticated person is inclined to think that because

> he draws water from a well by the bucketful, that means that water exists naturally in bucket-sized units. Water is a continuous fluid, and it is we who divide it when we scoop it up. The man-water interface, as we might call it, doesn't sanction inferences to the nature of water as such. Why then should we be so impressed by Millikan when he makes the corresponding move from the oil-electricity interface to the nature of electricity as such? ... The ultimate cause of our acceptance is that a sufficient number of trusted authorities can be found who are prepared to let such an inference pass muster, or even encourage it, even though in other circumstances (e.g., standing round the village well) they would laugh at it. (Barnes et al., 1996, pp. 26–27, original italics, underlining added)

This quotation represents an interesting piece of thinking about scientific methodology. I would agree with the first part and would like to endorse it by citing from a methodologist, Campbell (1988):

> [T]he objectivity of physical science does *not* come from the fact that single experiments are done by reputable scientists according to scientific standards. It comes instead from a social process which can be called competitive cross-validation ... and from the fact that there are many independent decision makers capable of rerunning an experiment, at least in a theoretically essential form. The resulting dependability of reports ... comes from a social process rather than from dependence upon the honesty and competence of any single experimenter. (pp. 302–303, original italics)

Furthermore, Campbell conceptualizes the social aspect of science as a systematic norm of distrust (organized skepticism), which facilitates validity by peer monitoring. It is plausible to suggest that "trusted authorities" (Barnes et al., 1996) and "independent decision makers" (Campbell, 1988) perhaps play the same role in scientific methodology.

In the second part, Barnes et al. (1996) state that "water is a continuous fluid" and that the "man–water interface" does not sanction inferences with respect to the nature of water. Precisely, it is the atomic theory that shows that based on the particulate nature of matter, "water is not a continuous fluid." Understanding particulate nature of matter is counterintuitive and no wonder that high school and even college students have considerable difficulty (cf. Gabel & Bunce, 1994). As a thought experiment, one could even argue that water could be scooped from a well as bucket-sized units, if the size of the bucket was adjusted to contain exactly 2.988 $\times 10^{-23}$ g (one molecule) of water. However, the important point is that both the particulate nature of matter and the elementary electrical charge are sanctioned by the "village elders" meeting around the well. Generally, the decisions taken at the village well, both in the past and at present, are well corroborated by experience over a period of time. My contention is that the decisions taken by the village elders are validated by peer monitoring. In the case of the elementary electrical charge the village well meeting could have been the 79 Meeting of the British Association for Advancement of Science held in Winnipeg, Canada, August 25–September 1, 1909. Holton (1978) has already drawn attention to the importance of this meeting, at which the presidential address was delivered by J.J. Thomson and E. Rutherford addressed the Mathematical and Physical Science Section. Rutherford emphasized the role played by atomic theory and traced its origin to Dalton's work in 1805, followed by the contributions of J.C. Maxwell and R. Clausius, "Brownian movement," Perrin, and finally sets the stage for what at that time was paramount:

> We have referred earlier in the paper to the work of Ehrenhaft on the Brownian movement in air shown by ultra-microscopic dust of silver. In a recent paper [1909] he has shown that each of these dust particles carries a positive or negative charge. The size of each particle was measured by ultra-microscope, and also by the rate of fall under gravity. The charge carried by each particle was deduced from the measured mass of the particle, and its rate of movement in an electric field. The mean value of *e* was found to be 4.6×10^{-10}. (Rutherford, 1909, p. 380)

In the next paragraph, Rutherford compared Ehrenhaft's value of *e* to his own (4.65×10^{-10}) and considered it to be of "considerable confidence." I wonder if the "village elders" could provide greater insight as to what was at stake. Robert Millikan was of course at the meeting and imbibed the lesson completely. Ehrenhaft was not at the meeting and nor did he see the "writing on the wall."

The role played by Ehrenhaft was, however, complex. After 1909, he interpreted his data to show that there was little evidence for the elementary electrical charge. Holton (1978) has shown how Ehrenhaft was greatly influenced by the anti-atomist and empiricist philosophy of E. Mach and W. Ostwald. Anti-atomists' ideas were still strong in 1910 and it is no surprise that both Millikan and Ehrenhaft played their part in this ongoing debate. Brush (1976) has presented this conflict succinctly:

> The leaders of this reaction [against the atomic theory], in the physical sciences were Ernst Mach, Wilhelm Ostwald, Pierre Duhem, and Georg Helm. Mach recognized that atomic hypotheses could be useful in science but insisted, even as late as 1912, that atoms must not be considered to have a real existence. Ostwald, Duhem, and Helm, on the other hand, wanted to replace atomic theories by 'Energetics' (a generalized thermodynamics); they denied that kinetic theories had any value at all, even as hypotheses. (p. 245)

Wilhelm Ostwald wielded considerable influence in scientific circles and in 1904 was invited by the Royal Society itself to deliver the Faraday Lecture and this is how a historian has summarized Ostwald's (enfant terrible of physical chemistry) presentation:

> Ostwald came to London and gave a stunning performance. It was not a talk aimed at convincing his audience. It was a talk aimed at crushing his audience. Ostensibly, his purpose was to show that all the laws of chemistry that could be deduced from the atomic hypothesis could equally well be deduced from the theory of chemical dynamics, which he told them was the most significant achievement of modern chemistry. … One could almost feel the audience fuming, prevented by etiquette from shouting against the sacrilege so unashamedly perpetrated by Ostwald at the very heart of London, at the Lecture Theatre, in fact, of the Royal Institution, where so much had been said about the real atoms all these past years. (Gavroglu, 2000, pp. 184–185)

We have been witness to two village well conversations, one at Winnipeg, Canada, in 1909 and the other at London in 1904, and in both leading village elders played an important role. This was the philosophical milieu that pervaded in the first decade of the twentieth century. Based on this scenario it is plausible to suggest that both Millikan and Ehrenhaft were strongly committed to the two leading philosophical currents of the time, and based on that presented plausible hypotheses. Ehrenhaft's data could be explained by an alternative rival hypothesis which was equally plausible, viz., the anti-atomist research program of Mach and Ostwald. In a sense the Millikan–Ehrenhaft controversy perhaps even represents an integral part of scientific development. Campbell (1988) emphasizes the rivalry and even proliferation of plausible hypotheses in both the social and the natural sciences. Similarly, Lakatos ([1970]) is quite emphatic: "[P]roliferation of theories cannot wait until the accepted theories are 'refuted' (or until their protagonists get into a Kuhnian crisis of confidence)" (p. 122).

Barnes et al. (1996) have raised another important issue, which I would like to develop further: "[W]e should avoid the inference to the rightness of Millikan's theory from the fact that it works" (p. 30). Based on Millikan's notebooks and data reduction procedures it is quite clear that his theory was not right even for some of his own data. Scientific theories need not be evaluated on the basis of rightness or wrongness, but instead on their heuristic power. Furthermore, according to Lakatos (1970) scientific theories are tentative. For example, was Thomson's (1897) theory right then and became wrong after Rutherford (1911) published his model of the atom? Rutherford's model increased the heuristic power of the theory, just as Millikan's determination of the elementary electrical charge increased the heuristic power of the atomic theory even further. Lakatos (1970, p. 123) attributes this conflation between "proven truth" and "heuristic power" to both W. Whewell and P. Duhem.

On the whole, Barnes et al. (1996) have facilitated our understanding of the experiment and thus played a role envisaged by Fuller (2000): "[S]ociologists can step into the breach when philosophers [Holton and Franklin in this case] cannot decide among themselves which methodology best explains a certain historical episode of scientific theory choice" (p. 141). Furthermore, Kitcher (2000) has cautioned: "Philosophical attempts to make the ultimately triumphant position rationally preferable even at early stages in the controversy seem to be doubly unfortunate" (p. 27). Similarly, Machamer et al. (2000, p. 16) have recommended that it would be advisable that philosophers, historians, and sociologists do not neglect each other's work.

An Appraisal of Goodstein's Interpretation

Of the five interpretations included in this study, the one by Goodstein (2001), who also had access to the Millikan's archives, was explicitly conducted to defend Millikan's methodology:

> What scientist familiar with the vagaries of *cutting-edge experimental work* would fault Millikan for picking out what he considered to be his most dependable measurements in order to arrive at the most accurate possible result? (Goodstein, 2001, p. 57, italics added)

Goodstein rightly recognizes that in "cutting-edge experimental work" scientists do resort to picking out data in order to arrive at the most accurate possible result. This raises two questions: (a) How did Millikan know in 1913 what was "the most accurate possible result"; (b) Millikan's "picking out" of data did not alter the right or the most accurate possible result. Goodstein leaves the first question unanswered and for the second presents Franklin's treatment of the original data which showed that had Millikan published data from all the drops available, this would not have changed the final value of *e*. This leads us to yet another question: Did Millikan perform the sort of analyses conducted by Franklin (1981) on published and unpublished drops, in order to justify the picking out of data. Apparently, based on the interpretations of Holton (1978), Franklin (1981), and Goodstein (2001) – all three consulted/checked Millikan's original notebooks – we have no evidence to the effect that Millikan could have performed Franklin-style analyses before deciding which drops to discard. Interestingly, Franklin (1997) implicitly does respond to this question by acknowledging:

> I speculate that this exclusion [drop of 16 April 1912 that gave a value of *e*, 40% low] was simply to avoid giving Felix Ehrenhaft ammunition in the charge-quantization controversy. Later analysis has shown that the data for this drop were indeed unreliable. (p. 28) [For later analysis Franklin refers the reader to: Fairbank & Franklin, 1982]

Given the importance of this drop of April 16, 1912 (Franklin, 1981, p. 200, considered it to be the most worrisome), one would have expected a detailed treatment in later analyses. Surprisingly, however, Fairbank and Franklin (1982) make no mention to this drop. Similarly, a later publication (Franklin, 1986) provides no further information with respect to this drop. In other words, given the controversy with Ehrenhaft in 1913, Millikan discarded data without performing the sort of analyses conducted later by Franklin (1981) and Fairbank and Franklin (1982). One could go further and speculate: If Millikan had conducted such analyses, it could have provided him with a good defense for excluding drops and he most probably would have included it in his publication (Millikan, 1913). This, of course, leads to the big question: In 1913, there was no way for Millikan to have known with certainty as to which was the "expected correct" value for *e*. His best guides were his presuppositions and the advice of the village elders (Rutherford & Geiger, 1908).

Goodstein (2001) is at pains to justify/explain the following problematic statement from Millikan (1913): "It is to be remarked, too, that this is not a selected group of drops but represents all of the drops experimented upon during 60 consecutive days" (p. 138). Goodstein (2001) cites this statement thrice in his paper with the following comments:

> The question is why did Millikan mar his masterpiece (Millikan, 1913) with a statement that is clearly false. (p. 57)

> Still, Millikan's paper contains that nagging and blatantly false … (p. 58).

> … [that] damning remark … (p. 59).

Finally, in an attempt to diminish the significance of the statement, Goodstein (2001) resorts to a truly tautological argument:

> So the damning remark is made, not about whether charge comes in units or what the value of *e* is, but in regard to getting the correction to Stokes's law right. Millikan is merely saying here that all of the 58 drops he just discussed confirm his presumed formula for amending Stokes's law. (p. 59)

Based on this presumption, Goodstein (2001) goes on to conclude: "Thus a careful reading of the context of Millikan's words greatly diminishes their apparent significance as evidence of misconduct" (p. 59). This raises an important issue: Millikan could easily have reported that based on his correction of Stokes' law, only 58 drops qualified for the study. He did not have to mention that these 58 drops formed part of a bigger group of drops. On the contrary, in order to make his case stronger (given the challenge from Ehrenhaft) he stated emphatically: "*[T]his is not a selected group of drops but represents all of the drops experimented upon*" (p. 138, original italics).

In order to diminish the significance of Millikan's "misconduct" Goodstein (2001) reproduces the following passage from Millikan (1913):

> Table XX, contains a *complete* summary of the results obtained on all of the 58 different drops upon which complete series of observations like the above were made during a period of 60 consecutive days. (Goodstein, 2001, p. 58; Millikan, 1913, p. 133, italics added)

Goodstein comments on this assertion of Millikan (1913) in the following terms: "Millikan didn't detail why he had not considered his evaluation of some drops to be sufficiently complete. ... The clear implication of this statement in his paper is that there *were* drops for which the data were not complete enough to be included in the analysis" (Goodstein, 2001, pp. 58–59, italics in original, underlining added).

Many years later Millikan (1947) recounted his experiments reported in Millikan (1913) in much the same way as in the original. There are, however, some significant changes as can be observed from the following (cf. with the quote from Millikan, 1913, p. 133, cited above):

> The numerical data from which these curves are plotted are given *fairly fully* in Table IX [Table XX in Millikan 1913]. It will be seen that this series of observations embraces a study of 58 drops. These drops represent all of those studied for 60 consecutive days, *no single one being omitted.* (Millikan, 1947, pp. 108–111, italics added. The quotation starts on p. 108 and finishes on p. 111, pp. 109–110 being devoted to Table IX, reproduced exactly as in Table XX of Millikan 1913)

It can be observed that the word *complete* [important for Goodstein] has been replaced by "*fairly fully*" and the phrase "*no single one being omitted*" has been added in Millikan (1947). How do we interpret these changes? Apparently, Millikan is alluding to the fact that the data presented in Table XX/IX is after all not *complete* but "fairly *fully* [complete]." Furthermore, the reference to "*no single one being omitted*" is obviously more categorical and hence makes the explanation for having excluded drops even more problematic and thus the evidence for "misconduct." Indeed, if we try to understand Millikan's handling of data with no reference to his presuppositions (Holton, 1978) or "hard-core" of his research program (Lakatos, 1970), then some degree of "misconduct" can be perceived.

A Crucial Test: The Second Drop (Reading) of 15 March 1912

This drop can provide further insight into Millikan's handling of the data as it formed part of the handwritten notebooks and studied within the period in which Franklin considered Millikan's apparatus to be working well. Holton (1978) devotes almost a page (pp. 209–210) to analyze Millikan's handling of this drop, and also reproduces the corresponding page from the handwritten notebooks (Fig. 4, p. 207). This was a heavy drop and it did not give the results expected by Millikan, which led him to note in his notebook, "*Error high* will not use" (italics by Millikan, reproduced in Holton, 1978, p. 209). About 3 cm below this Millikan added: "Can work this up & probably is ok but point is [?] not important. Will work if have time Aug. 22" (reproduced in Holton, 1978, p. 209). Holton considered that the latter part was added "probably later" (p. 209). It is plausible to suggest that guided by his presuppositions Millikan did not waste time in investigating the reason for the high error and instead went on to study another drop.

Now let us see how Franklin (1981), who also reproduced the corresponding page from the notebooks (Fig. 1, p. 188), handled this drop. Franklin pointed out that the exclusion of this drop had 'bothered Holton' (p. 195). After providing some details about the results, Franklin (1981) concluded:

> Although there is a substantial difference between the results of the two methods of calculation, larger than any for which Millikan gave data in his published paper, it is no larger than the difference in some events that Millikan published, as shown in the laboratory notebooks. There is no reason to assume, as Holton seems to do, there were unstated and perhaps unknown experimental reasons for the exclusion of this event. (p. 195)

No further arguments or analyses are presented as to why the drop was excluded or as to what could have bothered Holton. Let us go back to Holton (1978) to see how he interpreted the data with respect to this drop:

> [T]he entries on the right-hand page [Figure 4, p. 207], which Millikan abandoned, make excellent sense if one assumes that the smallest charge involved is not *e* but, say, one tenth *e*. ... From Ehrenhaft's point of view, it is the assumption of integral multiples of *e* that forces one to assume further, without proof, a high 'error' to be present and thus leads one to the silent dismissal of such readings and hence of the possibility that the quantum of electric charge may be 0.1*e*. (p. 210)

This clearly shows the role of presuppositions in the interpretation of data. For Millikan this drop simply represented a case of "high error" and according to Franklin there were other drops with similar characteristics. On the other hand, for Ehrenhaft this drop provides evidence for the existence of fractional charges (0.1*e*). Readers would have liked to see Franklin's detailed analysis with respect to this dilemma, and thus provide a rebuttal to Holton's interpretation.

Surprisingly, Goodstein (2001) makes no attempt to analyze Holton's (1978) interpretation of this drop. Barnes et al. (1996) have rightly recognized the importance of this drop:

> The point at which one would expect the import of Franklin's criticism to stand out with maximum clarity would be in his handling of the very example on which Holton's case crucially depends – the discarded second reading of 15 March 1912, where Millikan appeared to have captured a sub-electron charge *e*/10. What did Franklin make of this? ... Nothing Franklin says shows Holton is wrong. (pp. 44–45)

Finally, it is instructive to go back once again to Holton (1978) to understand the dilemma: "In Millikan's terms, on the contrary, such an interpretation of the raw readings would force one to turn one's back on *a basic fact of nature* – the integral character of *e* – which clearly beckoned" (p. 210, italics added). Could "a basic fact of nature" here represent the "presuppositions." The crux of the issue is that Holton's interpretation based on Millikan's methodology of defending his "presuppositions" has not been rebutted in subsequent interpretations. In other words, there was an alternative framework

(Ehrenhaft's) based on a different set of suppositions, viz., quantum of electric charge could have been 0.1*e*, which could have explained the data.

An Appraisal of Hintikka's (2005) Interpretation

Hintikka (2005) cites Babbage's classic taxonomy of scientific fraud, who distinguishes three kinds of such fraud: (a) forging; (b) cooking, and (c) trimming (Babbage, 1830). Many prominent scientists have been accused of omitting data, including a figure as prominent as Isaac Newton, who maintained the impossibility of an achromatic lens. The most thoroughly analyzed case is of course that of Millikan's oil drop experiment. According to Hintikka, omitting data is a clear-cut case of cooking, and if one follows the traditional approach to scientific method (depicts a step from particular data to a generalization or law), omitting data does indeed amount to some degree of suspicion. So, what is Hintikka's solution for this impasse?

Hintikka's Interrogative Model of Inquiry

The basic idea behind this model is that scientific experiments are considered as questions put to nature, and the scientific inquiry as a process based on: (a) definitory rules; and (b) strategic rules. In general, whenever the presupposition of a question has been established, the inquirer may formulate the question. If an answer is forthcoming, it is added to the list of available premises, which helps to place the data omission problem in a perspective. In the case of scientific inquiry (e.g., oil drop experiment), typical answers are outcomes of observations and experiments. In order to illustrate his model of interrogative inquiry, Hintikka draws an analogy with the cross-examination of a witness in a court of law, and suggests that the inquirer cannot simply accept all answers at their face value, as some of them may be false. At this stage the inquirer must test the veracity of the answers by further questions and answers, based on strategic rules for rejection or what Hintikka (2005) refers to as "bracketing of an answer" (p. 171). For more details on this interrogative model of inquiry see Hintikka (1999).

Hintikka on Millikan's Procedure

According to Hintikka, a witness's testimony in a court of law is tested by comparing it with other kinds of testimony and physical evidence. Based on this analogy, Hintikka (2005) has argued cogently that "[t]he same holds of course for scientific inquiry: the 'answer' (result) yielded by one experiment is tested by repeating the same experiment or by performing a collateral one" (p. 172). This analysis provides an interesting insight: The decision to omit or not to omit data is a strategic decision, whereas, in contrast, the definitory rule only provides guideline for what a scientist must do or not do. In other words, if a scientist has reasons to think that something has gone wrong in an experiment, he has the right to disregard the result.

At this stage, Hintikka (2005, p. 172) specifically refers to the five drops for which Millikan calculated the value of *e* and still rejected them as their values did not agree with his expectations. Franklin (1981, p. 195) had considered this a "clear case of trimming." Hintikka (2005), in contrast, considered this as "a beautiful case of the strategic nature of data omission. What was the strategic situation that Millikan was in? He apparently had strong theoretical reasons to think that the value of *e* was a constant. … In view of his theoretical background, the fact that the result differs from the others is ipso facto a danger sign. It suggests that there might be something wrong with the exceptional observation. This is a perfectly rational suggestion" (p. 173). Traditionally, omitting data is automatically considered to be against the norms of the scientific method (this precisely was Ehrenhaft's strategy, see earlier discussion in this chapter). Hintikka, on the contrary not only justifies omitting ("bracketing") data as a strategic decision, but rather goes beyond by suggesting that "[c]ertainly an injunction never to omit data cannot be part of any absolute ethics of science" (p. 172). Interestingly, Hintikka suggests that just as a scientist may resort to "bracketing" of data, he may also allow "unbracketing" at some later stage of inquiry.

Hintikka on the Drop of April 16, 1912

This drop was considered by Millikan to be one of his very best observations and at one stage he wanted to publish it. However, this drop gave a value of *e* about 40% lower than what he had expected. Millikan finally dismissed the event with the comment "Won't work" and did not publish it. This event was considered by Franklin (1981, pp. 200–211) to be a serious omission of data. According to Hintikka, based on the traditional view, this looked like a "barefaced case of cooking" (p. 174). Hintikka goes into considerable detail in order to analyze this drop and considers it to be yet another interesting strategic decision. He starts by asking: What was the question that Millikan was putting to nature? and suggested two possible questions (that are not exclusive, p. 174):

(a) What is the value of *e*?
(b) Does electric charge always come in multiples of one and the same charge?

If Millikan had asked question (a), then he was justified in being suspicious of this observation (drop of April 16, 1912). Here Hintikka indulges in virtual history by imagining that one of Millikan's oil drops had yielded a value of *e*, about twice his other results. Millikan' dilemma would have been: Was this a valid observation and hence needed to be included? Knowing Millikan's research question (a), he would undoubtedly have concluded that the drop in question had two electrons in place of one. On the contrary, if Millikan were trying to prove that electric charge is quantized and assumes only values that are multiples of *e* (see question b above), then the observation of the odd fractional charge could have been a potential counterexample, and its omission problematic. In a personal communication to the author, Hintikka (2006) expressed this even more clearly: "Moreover, what is the *presupposition* that Millikan had to make? He did not have to assume that electric charge is always a multiple of the same constant. It would have sufficed for his experiment to assume that in a number of cases, apparently most of them, the charge is one and the same. As long as Millikan was merely establishing that value, his omission procedure could be defended" (emphasis added).

Finally, Hintikka's analysis (interrogative model of inquiry) of Millikan's oil drop experiment provides not only an understanding that experimental data have to be validated by alternatives and that the injunction "never to omit data" cannot be part of any absolute ethics of science. Interestingly, he goes even beyond by suggesting: "I have been amazed by the poor grasp of too many writers of the most fundamental aspect of scientific inquiry. In my mind, it is a much worse breach of the ethics of science to accuse Millikan of fraud merely because he omitted data than anything he did himself" (Hintikka, 2006).

Educational Implications of the Historical Reconstruction of the Oil Drop Experiment

Millikan–Ehrenhaft Controversy

Given the controversial nature of the experiments and considerable discussion among historians and philosopher of science, one would expect that science educators would take notice of such developments. Niaz (2000b) found that of the 31 general chemistry textbooks analyzed none mentioned the controversy. Similarly, Rodríguez and Niaz (2004c) have reported that of the 43 general physics textbooks analyzed none mentioned the controversy. Some textbooks explicitly denied that the drops studied by Millikan had fractional charges, i.e., a charge unequal to an integer times the electron charge and following is an example: "Millikan found that each droplet had a charge of either zero, *e*, 2*e*, 3*e*, 4*e*, or some whole number times *e*, but never a fraction of *e*. …No one has ever found an electric charge that is not a whole number times the elementary charge, the charge on the electron" (Hulsizer & Lazarus, 1977, p. 227). This shows the degree to which textbooks embellish the historical record in order to present students a "pristine" vision of the scientific endeavor.

Millikan's Guiding Assumptions

Based on the atomic nature of electricity, Millikan hypothesized the existence of an elementary electrical charge. Millikan found drops with a wide range of electrical charges. Despite such anomalous data, if it were not for the guiding assumption, Millikan would have abandoned the search for the elementary electrical charge. Niaz (2000b) has shown that of the 31 general chemistry textbooks analyzed only six made a simple mention. Rodríguez and Niaz (2004c) have reported that of the 43 general physics textbooks analyzed only two made a simple mention and following is an example: "By observing the motion of the hundreds of droplets with different charges on them, Millikan *uncovered the pattern he expected*: the charges were multiples of the smallest charge he measured" (Olenick, et al., 1985, p. 241, emphasis added). "The pattern" can be considered as an oblique reference to his guiding assumption. The same textbook reproduced the following quote from Millikan's laboratory notebook: "One of the best ever [data] … almost exactly *right*. Beauty – publish." After reproducing the quote, authors asked a very thought-provoking question: "What's going on here? How can it be right if he's supposed to be measuring something he doesn't *know?* One might expect him to publish everything!" (p. 244). This is the closest that any textbook came with respect to mentioning Millikan's guiding assumption, and this particular textbook did not take advantage of presenting students an understanding of how scientists work and have presuppositions.

Dependence of the Elementary Electrical Charge on Experimental Variables

Millikan was constantly trying to improve his experimental conditions to obtain the charge on the droplets as an integral multiple of the elementary electrical charge. Some of the variables that he constantly referred to were evaporation, sphericity, and radius of the droplets, change in density of the droplets, changes in battery voltages, temperature, and viscosity of the air. Niaz (2000b) has shown that of the 31 general chemistry textbooks analyzed only two presented this heuristic principle satisfactorily and following is an example:

> Using an atomizer, microscopic spherical drops of oil are introduced into the space above two charged plates. Oil is used because it does not noticeably evaporate. … Gravity causes the drops to fall, but they are slowed by friction, due to the viscosity of the air. … The charge on the drop can be calculated from the value of its downward velocity, the magnitude of the potential difference, the known acceleration of gravity, the density of oil, and the air viscosity. (Segal, 1989, pp. 412–413)

Without the description of these and other experimental variables that made the oil drop experiment so difficult to perform, students are left with the impression that in order to discover new things, scientists need only to walk into the laboratory. Similarly, for general physics textbooks, Rodríguez and Niaz (2004c) have reported that only one textbook made a simple mention.

Millikan's Experiments as Part of a Progressive Sequence of Heuristic Principles

Millikan's work started by repeating and making a critical evaluation of the experimental work of Townsend, Thomson, and Wilson on charged clouds of water droplets.

Each stage in this historical process was characterized by guiding assumptions, improvement in experimental techniques, criticisms, and rebuttals. Niaz (2000b) has shown that of the 31 general chemistry textbooks analyzed only one (Burns, 1996) made a brief mention. Rodríguez and Niaz (2004c) have reported that of the 43 general physics textbooks analyzed, only one (Ohanian, 1987) described satisfactorily that Millikan's work formed part of a sequence of heuristic principles. Another textbook (Eisberg, 1973) made a simple mention.

Oil Drop Experiment in the Undergraduate Physics Laboratory

In an effort to explore further the presentation of the Millikan oil drop experiment, Rodríguez and Niaz (2004c) also analyzed 11 physics freshman laboratory manuals. Surprisingly, even the manuals do not deal satisfactorily with the experimental variables (used in the laboratory), which made the experiment so difficult. Just as textbooks, laboratory manuals attribute Millikan's success to his precise measurements and thus ignore the theoretical rationale behind his experimental data.

Kruglak (1972) specifically referred to the difficulties involved in the experiment and considered it to be one of the most frustrating in the undergraduate laboratory. Nevertheless, he also recognized that the oil drop experiment facilitated empathy for the challenges and vicissitudes of the physicist. Kapusta (1975) found that some of the drops have much higher velocities than others, and so should one choose to observe the fast drops, the slow drops, or those of intermediate velocities. Most teachers ignore the fact that this precisely was Millikan's dilemma, and a perspective based on history and philosophy of science (HPS) can help to resolve it (cf. Holton, 1978 for a particularly lucid and convincing account of this dilemma). In most traditional laboratory courses students are systematically misled into believing that the results from the oil drop experiment were straight forward and uncontroversial.

In a recent study, Klassen (2009) used an HPS perspective to perform the experiment with undergraduate physics students. In this laboratory strategy students were not left simply to deduce the fact that their data choice is guided by their presuppositions (quantization of electric charge) but rather directed with appropriate questions that challenged them to be reflective about Millikan's work. With this helpful background and guidance, students attained insight and better understanding of the underlying issues, as can be seen from the following response from one of the students:

> By having a preconceived notion of what *e* should be, we knew what to expect and disregarded observations that were not expected. In saying so, I believe Millikan depended on his preconceived notion as much as we did. It is likely that when Millikan noticed a quantization trend of the charges, he selected only those drops that would illustrate the phenomena and excluded those few that distort it. By doing so, he was able to illustrate his discoveries for us to understand undoubtedly. Had he not, who knows when we would finally acknowledge charge quantization? (p. 12)

This clearly shows that given the opportunity to think, reflect, and discuss the historical background of the experiment can facilitate a better understanding of science in the making.

Oil Drop Experiment in Advanced Courses

It could be argued that the oil drop experiment is too complex and cannot be presented in introductory freshman textbooks. So we decided to study physical chemistry textbooks, which are used in an advanced course that includes atomic structure as a topic. Niaz and Rodríguez (2005) reported that none of the 28 physical chemistry textbooks recognized the controversial nature of the oil drop experiment and ignored the heuristic principles. It is understandable that physical chemistry textbooks may not present the Millikan–Ehrenhaft controversy (it pertains to HPS?) but how do we justify that the experiment could be presented without a satisfactory explanation of the incidence of the experimental variables that made the experiment difficult. This is all the more paradoxical, as physical science is considered by its practitioners to be primarily an experimental science.

Conclusion: Is Closure Possible?

It is almost 90 years since most of the experimental work that led to the determination of the elementary electrical charge was conducted. Since then there has been considerable controversy amongst physicists, historians, philosophers, and sociologists as to how data was collected and the reduction procedures employed. Three of the researchers have had access to Millikan's handwritten notebooks. Given these circumstances, it is plausible to attempt a closure. In order to facilitate the closure, let us consider some hypothetical questions that are discussed in this section.

If Millikan's handwritten notebooks had been lost (or for that matter if Holton, Franklin, and Goodstein had not consulted them), the oil drop experiment would have remained an enigma for some (including Ehrenhaft), while for others it would have been a classical test case of how experiments unambiguously lead to theoretical formulations. Availability of Ehrenhaft's notebooks (which were lost when Nazi Germany took over Austria during the Second World War) could also have helped to provide further insight into our understanding of the experiment.

What would have been the consequences if the scientific community had recognized Ehrenhaft's findings and not Millikan's? This leads to a corollary, viz., what would have been the consequences if Millikan's formulation of the elementary electrical charge could not have been sustained by further experimental evidence. This issue is crucial as there is evidence to show that the choice between Millikan and Ehrenhaft was not automatic. Release of the Nobel Prize Archives shows that

> although Millikan was nominated for the prize in physics regularly from 1916 on, Svante Arrhenius, in the report he prepared on Millikan's work for the deliberations of the committee, noted as late as 1920 that even though most physicists had come to agree with Millikan in the dispute with Ehrenhaft, the matter was not yet resolved, and that Millikan should therefore not then be recommended for the prize. (Holton, 1988, p. 196)

Millikan, finally got the Nobel Prize in 1923. This contrasts with Franklin's (1981, p. 191) claim that the question of charge quantization was already settled even before Millikan (1913) was published. A plausible course of events that could have happened if Millikan's findings in the long run had been proven untenable is provided by Olenick et al. (1985):

> He [Millikan] had a pretty clear idea of what the result ought to be – scientists almost always think they do when they set out to measure something … it's actually a powerful bias to get the result he wants, because you can be sure that when he got a result he liked, he didn't search as hard to see what went right. *But experiments must be done in that way. Without that kind of judgment, the journals would be full of mistakes, and we'd never get anywhere. So, then, what protects us from being misled by somebody whose 'judgment' leads to a wrong result? Mainly, it's the fact that someone else with a different prejudice can make another measurement. … Dispassionate, unbiased observation is supposed to be the hallmark of the scientific method. Don't believe everything you read. Science is a difficult and subtle business, and there is no method that assures success.* (p. 244, italics added. David Goodstein, one of the co-authors apparently changed his mind in Goodstein, 2001)

Millikan and Ehrenhaft would never have discussed their experimental findings in the context of "a different prejudice" and problematical nature of "dispassionate, unbiased observation." Although during the controversy at one stage Millikan came quite close: "That these same ions have one sort of charge when captured by a big drop and another sort when captured by a little drop is obviously *absurd*. … Such an assumption [existence of a whole range of fractional charges, as suggested by Ehrenhaft] is not only too *grotesque* for serious consideration but is directly contradicted by my experiments" (Millikan, 1916, p. 617, italics added). In other words, Millikan was trying to convince the reader by recognizing that experimental observations are important, but there is something even more important, viz., the presuppositions, and any data that go against them would appear to be "absurd" and even "grotesque," hence subelectrons could not exist.

Even philosophers of science would now recommend that in cutting-edge scientific research, collection of experimental data (Ehrenhaft's strategy) in itself may not lead to new insights:

> Our vision of reality, to which our sense of scientific beauty responds, must suggest to us the kind of questions that it should be reasonable and interesting to explore. It should recommend the kind of conceptions and empirical relations that are intrinsically plausible and which should therefore be upheld, *even when some evidence seems to contradict them*, and tell us also, … what empirical connections to reject as specious, even though there is evidence for them – evidence that we may as yet be unable to account for on any other assumption. (Polanyi, 1964, p. 135, italics added)

It is plausible to suggest that Ehrenhaft's methodology approximated to the traditional scientific method which did not allow him to discard "specious drops." Millikan, on the other hand, in his publications espoused the scientific method, but in private (cf. handwritten notebooks) was fully aware of the dilemma faced and was forced to select data in order to uphold his presuppositions. No wonder, according to Fuller (2000), "something called the 'scientific method' is at best a philosophical caricature of what science is really about" (p. 212). As an indicator of how scientific methodology has progressed it is remarkable that even physicists now recognize in public that:

> Choices in the design of speculative experiments [cutting-edge] usually cannot be made simply on the basis of reason. The experimenter usually has to base her or his decision partly on what feels right, partly on what technology they like, and partly on what aspects of the speculations [presuppositions] they like. (Perl & Lee, 1997, p. 699)

Interestingly, Martin Perl recipient of the 1995 Nobel Prize for Physics is at present working on Millikan-type experiments in order to isolate fractional charges (quarks).

A closure to the controversy with respect to the oil drop experiment is possible if we recognize that Millikan's data selection procedure depended primarily on his perseverance with his presuppositions, viz., the existence of the elementary electrical charge and its magnitude, based on previous studies. Franklin's (1981) finding that the selection of the drops did not change the value of the elementary electrical charge (*e*) but only its statistical error carries little weight as Millikan did not perform Franklin-style analyses that could have justified the exclusion of drops. Furthermore, acceptance of Franklin and Goodstein's arguments approximates quite closely to what Kitcher (2000) has critiqued: "The most primitive type of rationalism proposes that scientific controversies are resolved by designing and carrying out crucial experiments" (p. 21). Similarly, Hentschel (2004, p. 609) has critiqued Franklin's exaggerated claims and suggested that despite occasional differences between public and private data (Millikan's case being an example), between what is published and what is recorded in the laboratory notebook, science remains a rational enterprise. At this stage it is important to recognize that such conflicts between interpretation of data and even ethical concerns form an inherent part (perhaps irresolvable for some) of the dynamics of scientific progress (cf. Segerstrale, 1995).

Millikan himself in his Nobel Prize speech provided a clue to his methodology, which has been generally ignored by historians, philosophers, and physicists:

> Indeed, *Nature* here was very kind. She left only a narrow range of field strengths within which such experiments as these are all possible. They demand that *the droplets be large enough so that Brownian movements are nearly negligible, that they be round and homogeneous, light and non-evaporable, that the distance be long enough to make the timing accurate, and that the field be strong enough to more than balance gravity by its pull on a drop carrying but one or two electrons*. Scarcely any other combination of dimensions, field strengths and materials, could have yielded the results obtained. (Millikan, 1965, pp. 57–58, italics added)

It is not far-fetched to suggest that Millikan's mention of "Nature" could very well mean "presuppositions" and the stringent experimental variables mentioned (given the difficulty of getting the right combination) inevitably lead to selection of drops. A recent attempt to study the structure of scientific controversies shows that the closure point cannot be fixed by logical or methodological rules and finally recognizes the role of "presuppositions": "What science says the world is like at a certain time is affected by the human ideas, choices, expectations, prejudices, beliefs, and assumptions holding at that time" (Machamer et al., 2000, p. 6).

Finally, it would be interesting to see what Martin Perl (Nobel Laureate, 1995), an experimental physicist actually involved in cutting-edge research related to the isolation of quarks, has to say with respect to Millikan's strategy with respect to the oil drop experiment: "I agree with your conclusion that Millikan believed that the only basic charge was that on the electron and he sorted his data to get the best measurement of that charge. I don't know what he would have done if there were say three different particles with different values of the electric charge"

(Perl, 2005). This may sound more convincing to many readers, especially those who do experimental work and consider it to be crucial in Millikan's success. Furthermore, Perl is very familiar with Millikan's experimental procedure as he is using an automated Millikan-style apparatus for the isolation of quarks (for details see, Perl & Lee, 1997; Perl et al., 2004).

This chapter shows the importance of the contributions of both Millikan and Ehrenhaft. Two brilliant, hard working and enthusiastic scientists who devoted their lives to follow single-mindedly the same problem, each in his own particular style. Crease (2003) refers to a "poignant moment" in the life of Ehrenhaft, when at the annual meeting of the American Physical Society in New York in 1946 he approached the podium demanding to be heard and was then politely escorted out of the room. Again, Crease (2003) recounts how on a "pilgrimage" to pay homage to the "defining moment of our electronic age" he visited the Ryerson Hall at the University of Chicago. To his surprise he found no commemorative plaque paying tribute to Millikan's great contribution, nor any trace of his experiment in the building. Such is the march of history and future generation of students interested in the dynamics of scientific progress will emit their own verdict.

Chapter 8
Paradox of the Photoelectric Effect: Einstein and Millikan

Introduction

It is almost 100 years since Einstein (1905) published his hypothesis of lightquanta to interpret the photoelectric effect. This hypothesis has been the subject of considerable research by historians (Brush, 2007; Klein, 1963; Kragh, 1999; Stuewer, 1970, 1975; Wheaton, 1978, 1983). Einstein proposed that ordinary light behaves as though it consists of a stream of independent localized units of energy that he called *lightquanta*. He was led to this revolutionary view by a statistical analysis of the properties of an idealized sample of radiation in thermodynamic equilibrium. His suggestion of this hypothesis arose from the close analogy he perceived between the behavior of radiation and the behavior of a gas (Wheaton, 1983, p. 106). According to Einstein, if light consists of localized quanta of energy, an electron in an atom will receive energy from only one lightquantum at a time. Monochromatic light of frequency v can therefore grant electrons only energy, hv, where h is Planck's constant. If some small part p of that energy must be used to release the electron from the metal itself, all electrons of charge e so released will be stopped by a decelerating potential P, following the relation:

$$\frac{1}{2}mv^2 = Pe = hv - P \qquad (8.1)$$

($\frac{1}{2}mv^2$ = maximum kinetic energy of the ejected electrons). Einstein's (1905) prediction that the stopping potential "should be a linear function of the frequency of the incident light, when plotted in Cartesian coordinates, and its slope should be independent of the nature of the substance investigated" (p. 146, English translation from Jammer, 1966, p. 35) became the cornerstone of Millikan's research program. Interestingly, Einstein's hypothesis also explained Lenard's (1902) triggering hypothesis, viz., the maximum velocity of photoelectrons must be independent of the intensity of light and the energy received by an electron depends on the frequency. According to Wheaton (1983):

> Einstein's hypothesis of lightquanta was not taken seriously by mathematically adept physicists for just over fifteen years. The reasons are clear. It seemed to be an unnecessary rejection of the highly verified classical theory of radiation. … How lightquanta could possibly explain interference phenomena was always the central objection. (p. 109)

M. Niaz, *Critical Appraisal of Physical Science as a Human Enterprise: Dynamics of Scientific Progress*, Science and Technology Education Library 36,

In the face of this criticism, many physicists (including Millikan) thought that Einstein would retract. Nevertheless, Pais (1982) has reported that "I find no evidence that he [Einstein] at any time withdrew any of his statements in 1905" (p. 383). A historical reconstruction of this episode in the history of science is presented in the following sections:

1. Millikan's experimental determination of Planck's constant (*h*)
2. Millikan's interpretation of Einstein's hypothesis of lightquanta
3. Millikan's interpretation of the hypothesis of lightquanta in retrospect
4. Holton's two ironies
5. The third irony
6. Conclusion

Millikan's Experimental Determination of Planck's Constant (*h*)

Starting around 1904, Millikan devoted considerable effort to the experimental determination of Einstein's photoelectric equation (see Eq. (8.1)) and as a consequence calculated the experimental value of Planck's constant *h* (Millikan, 1913a, 1914, 1915, 1916a, b). Millikan's value for *h* (6.57×10^{-27} erg. s) came quite close to that reported by Planck (6.55×10^{-27}). Millikan was quite satisfied with all the technical innovations that he developed for his experimental measurements and concluded: "so far as experiment has now gone, Einstein's photoelectric equation, whatever may be said of its origin, seems to stand up accurately under all of the tests to which it has been subjected" (Millikan, 1916a, p. 31). Later critical evaluation of Millikan's experimental determinations also recognized its importance: "Millikan's results were never questioned, and were quickly recognized by leading European physicists to be definitive. But acceptance of the Einstein law of the photoelectric effect did not carry with it acceptance of the hypothetical lightquantum" (Wheaton, 1983, p. 241).

Actually, it was possible to derive the experimentally confirmed equation without the lightquantum hypothesis, which thus constituted a truly novel prediction (Kragh, 1999, p. 67).

Millikan's Interpretation of Einstein's Hypothesis of Lightquanta

Given Millikan's penchant for controversy (cf. oil drop experiment, Holton, 1978; Niaz, 2000b, 2005b), it is interesting to note that in the same publication (Millikan, 1916b), on the one hand he recognized the validity of the Einstein photoelectric equation and on the other hand he questioned the underlying hypothesis of lightquanta: "Despite then the apparently complete success of the Einstein equation, the physical theory of which it was designed to be the symbolic expression is found so untenable that Einstein himself, I believe, no longer holds to it. But how else can the equation be obtained?" (Millikan, 1916b, p. 384). Despite underdetermination of theories by evidence (Duhem–Quine thesis, see Chapter 2), this may still sound incredible to any student of history and philosophy of science. A modern historian-chemist considered this as "the strongest kind of evidence for the validity of the quantum theory in general, and for the Einstein photon concept in particular" (Cropper, 1970, p. 28). This clearly shows that what we consider in retrospect to be strong evidence was not considered so by the original protagonists. It is plausible to suggest that the role of presuppositions (classical wave theory of light in this case) is crucial in the interpretation of data, especially in cutting-edge research.

Actually, Millikan went far beyond doubting further development in Einstein's ideas since 1905 (having Einstein restated his position in 1916), and it is worthwhile to follow Millikan's train of thought in some detail:

> It was in 1905 that Einstein made the first coupling of photo effects … with any form of quantum theory by bringing forward the bold, not to say *the reckless, hypothesis* of an electro-magnetic light corpuscle of energy $h\nu$, which was transferred upon absorption to an electron. This hypothesis may well be called reckless first because an electromagnetic disturbance which remains localized in space seems a violation of the very conception of an electromagnetic disturbance, and second because it flies in the face of the *thoroughly established facts of interference*. The hypothesis was apparently made solely because it furnished a ready explanation of one of the most remarkable facts brought to light by recent investigations, viz., that the energy with which an electron is thrown out of a metal by ultra-violet light or X-rays is independent of the intensity of the light while it depends on its frequency. This fact alone seems to demand *some modification of classical theory* or, at any rate, it has *not yet been interpreted satisfactorily in terms of classical theory*. (Millikan, 1916b, p. 355, italics added)

It is not far-fetched to suggest that Millikan's train of thought could be understood along the following lines: We have thoroughly established facts of interference based on the classical wave theory → although, some modification of classical theory may be necessary → nevertheless, Einstein's hypothesis of lightquanta is reckless. Later, more evidence will be provided to substantiate this suggestion.

Millikan was awarded the 1923 Nobel Prize in physics for his contributions to the experimental determination of both the elementary electrical charge and Planck's constant, based on Einstein's photoelectric equation. In his acceptance speed delivered on May 23, 1924, Millikan recounted the difficulties during the experiments and finally concluded on a pessimistic note:

> [T]he conception of *localized* light-quanta out of which Einstein got his equation must still be regarded as far from being established … until it [hypothesis of lightquanta] can account for the facts of interference and the other effects which have seemed thus far to be irreconcilable with it, we must withhold our full assent. Possibly the recent steps taken by Duane, Compton, Epstein and Ehrenfest may ultimately bear fruit in bringing even interference under the control of localized light-quanta. *But as yet the path is dark*. (Millikan, 1965, pp. 63 and 65)

This shows that even in 1924 Millikan did not accept Einstein's theory of lightquanta. Millikan was, of course, not alone in his rejection of Einstein's theory. According to Stuewer (1975):

> Therefore, in common with Lorentz, Thomson, Sommerfeld, and Richardson, Millikan did not take the validity of Einstein's *equation* – which he established beyond doubt – to constitute proof of the validity of Einstein's light quantum *hypothesis*. (p. 75, original italics)

Furthermore, acceptance of Einstein's theory had to compete with rival theories put forward by J.J. Thomson, A. Sommerfeld, and O.W. Richardson between 1913 and 1916 (Stuewer, 1970).

Millikan's Interpretation of the Hypothesis of Lightquanta in Retrospect

In a recent study, Holton (1999) has raised an important issue: "So Millikan's [1916b] paper is not at all, as we might now naturally consider it to be, an experimental proof of the quantum theory of light" (p. 234). This statement has important implications. The same publication (Millikan, 1916b) was considered by its own author as an experimental test of Einstein's photoelectric equation and in no way a confirmation of the underlying hypothesis of lightquanta. On the contrary, most textbooks and physicists at present consider Millikan's (1916b) experimental data as evidence for Einstein's quantum theory (hypothesis of lightquanta). To make things even more difficult, some scholars even consider that Einstein and not Planck was the originator of the quantum hypothesis (Brush, 2000; Kragh, 1999; Kuhn, 1978; Wheaton, 1983). How to resolve this dilemma? A plausible explanation is provided by Millikan's commitment to his presuppositions with respect to the wave theory of light, stated quite explicitly in many articles (Millikan, 1916a, b, 1965). The degree to which Millikan adhered to the wave theory of light as a presupposition is corroborated by the following scholars:

(a) According to Wheaton (1983): "Despite his uncontested proof of Einstein's linear photo effect relation, Millikan never doubted, until 1922, that an explanation of the result would be found without the suspect lightquantum. His particularly strong opposition may be attributed both to his respect for European mathematical physics – he had studied under Planck and Nernst in Berlin – and to the tradition that his teacher, Albert Michelson, had developed in American precision measurement in wave optics" (pp. 258–259).
(b) Chang (1995) has presented the thesis of "Provisional phenomenalism" in order to understand scientific change, viz., if observation is theory-laden, instability in theory very likely means instability in observation. Under these circumstances this thesis suggests that scientists "can do experiments on the basis of well-established phenomenological regularities, which are not likely to change no matter what happens in the realm of high-level theory" (p. 134). After studying various cases in the history of science (including Millikan's photoelectric experiments), Chang found that this is what most scientists did, which led him to conclude that in the case of quantum-physical experiments, "experimenters relied on classical

reasoning in almost all experiments investigating or utilizing the 'particle-like' aspects of microscopic objects: location, velocity, acceleration, mass, electric charge, etc. *The infiltration of quantum-physical experimentation by classical reasoning was deep and widespread*" (Chang, 1995, p. 125, emphasis added).

(c) Holton (1999) has described Millikan's presupposition in truly picturesque terms: "What we now refer to as the photon was, in Millikan's view, a 'bold, not to say the reckless, hypothesis' – reckless both because it was contrary to such classical concepts as light being a wave propagation phenomenon, and because of the 'facts of interference'. ... In the background we glimpse the presence of Michelson, the 'Artist of Light,' who was Millikan's admired patron and colleague at the Ryerson Laboratory, the 1907 Nobelist, famous for his interferometers, the work carried out with their aid – and for his adherence to ether physics to his death in 1931" (pp. 233–234).

Holton's Two Ironies

In his reconstruction of Millikan's struggle to understand the meaning of Planck's constant, Holton (1999) refers to two ironies:

(a) Holton (1999) refers to the fact that Millikan published his *Autobiography* in 1950 at the age of 82 and reproduces the following passage referring to experimental data obtained in his laboratory, with respect to Einstein's photoelectric equation, which "proved simply and irrefutably, I thought, *that the emitted electron that escapes with the energy hv gets that energy by the direct transfer of hv, units of energy from the light to the electron* and hence scarcely permits of any other interpretation than that which Einstein had originally suggested, namely that of the semi-corpuscular or photon theory of light itself" (Millikan, 1950, pp. 101–102, emphasis in the original by Millikan). Holton (1999) then refers to this as the First Irony in the following terms: "In the end, Millikan re-imagined the complex personal history of his splendid experiment to fit the simple story told in so many of our physics textbooks" (p. 236). Interestingly, Stuewer (1975, p. 88) considers this adjustment on the part of Millikan as "shocking," considering the fact that even in 1924, in his Nobel Prize acceptance speech, Millikan still questioned Einstein's hypothesis of lightquanta.

(b) The Second Irony that Holton (1999) refers to is with respect to some postmodern accounts of progress in science: "In the current deconstruction of science by non-scientists, much is made of the tendency of researchers to ask the questions and analyze results in the light of their own beliefs. Millikan's paper is an excellent counter-example, being clearly a case of someone doing superlative work in establishing a result from a theory in which he himself clearly had little faith. In this case at least, the carefully established facts spoke for themselves through a very skeptical scientist" (p. 236, Holton attributes this suggestion to Tudor Wyatt Johnston).

The Third Irony

Even Millikan with an excellent background and training in the determination of both *e* (elementary electrical charge, Millikan, 1913b) and *h* (Planck's constant, Millikan, 1916b) could not fathom beyond the "observable empirical data," and give up his prior presuppositions (cf. Chang, 1995; Holton, 1999; Wheaton, 1983) to at least consider the quantum theory on the same footing as wave theory, from say 1916 onwards. This shows that despite the importance of empirical data, presuppositions play a crucial role in a scientist's struggle in order, at times, to understand experimental data (Planck's constant, *h*) and on occasions even to discard data (*e*, elementary electrical charge).

Conclusion

The role of presuppositions in scientific progress has been recognized by philosophers of science in the following terms: "hard core" of a research program (Lakatos, 1970), guiding assumptions (Laudan et al., 1988), and presuppositions (Holton, 1978). According to Lakatos (1970) a scientist generally does not abandon his "hard core" in the face of anomalous data. This is exactly what Millikan did in the case of the oil drop experiment and it gave good dividends (cf. Holton, 1978). In his photoelectric experiments, Millikan followed the same strategy, that is a strong belief in the classical wave theory of light (presupposition). In this case, however, Millikan's perseverance with the presupposition did not bear fruit and he ultimately had to change the "hard core" of his research program.

The relationship between Millikan's two major research programs (oil drop experiment and photoelectric effect) and the role played by presuppositions has been explained by Holton (1998) in cogent terms:

> If Millikan's only scientific achievement was the oil drop experiment, he might be open to the charge that he was lucky in guessing at the usable data [right presupposition], or fortunate in his obstinacy. Such a charge would collapse in the face of his next and perhaps most influential work, the resumption of research on the photoelectric effect. Here he found himself working with the *wrong presupposition*, but he knew how to rid himself of it eventually. Millikan launched into that work with the same energy and obstinacy as into his earlier work on the quantization of the charge of the electron [oil drop experiment], yet with the opposite assumption. (p. 72, emphasis added)

This shows clearly that presuppositions (right or wrong) form an integral part of the research program of a scientist and that at times it requires considerable amount of intellectual effort to provide grounds for the "hard core" to crumble and make way for the new framework. Both Duhem (1914) and Lakatos (1970) have emphasized that the "hard core" (based on presuppositions) is not only difficult to refute but it also develops slowly and not like Athene from the head of Zeus. Millikan believed in the wave theory of light without any reservations (despite experimental evidence to the contrary) and was skeptical of the lightquanta (quantum hypothesis) as late as 1924. Holton's (1999) ironies provide an interesting historical backdrop to Millikan's attempt in his autobiography to "rewrite" history to make his account more in tune with the unfolding of historical events after 1924.

Chapter 9
Bending of Light in the 1919 Eclipse Experiments: Einstein and Eddington

Introduction

Britain's Royal Astronomical Society organized two expeditions to observe the 1919 eclipse of the sun in order to provide experimental evidence for Einstein's General Theory of Relativity. One of the parties (A.S. Eddington and E. Cottingham) was to observe the eclipse in Principe, an island off the coast of Africa, and the other (A. Crommelin and C. Davidson) in Sobral, Brazil. Complete report of the expeditions was published by Dyson et al. (1920). Nevertheless, a review of the literature shows that not only were the expeditions difficult to conduct, but the data produced also generated considerable amount of controversy even up to the present (Brush, 1989, 1999; Collins & Pinch 1993, 1998; Earman & Glymour, 1980; Hentschel, 1992; Hudson, 2003; Mayo, 2003; Moyer, 1979; Sachs, 1988; Sponsel, 2002; Stanley, 2003; Will, 1990).

Frank Watson Dyson, the Astronomer Royal, and Arthur Stanley Eddington played leading roles in the organization of the expeditions and later in the interpretation of the results, in order to provide support for Einstein's General Theory of Relativity. According to Dyson et al. (1920), "The purpose of the expeditions was to determine what effect, if any, is produced by a gravitational field on the path of a ray of light traversing it" (p. 291), and expected the following possible alternative results: (a) the path of a ray of light is uninfluenced by gravitation; (b) the energy or mass of light is subject to Newton's law of gravitation, which leads to a deflection of 0.87″; (c) the course of a ray of light is in accordance with Einstein's General Theory of Relativity, which leads to a deflection of 1.75″. The predictions of both Newton's and Einstein's values for deflection of light in the sun's gravitational field were well known in the literature before the expeditions were undertaken. Both Dyson and Eddington took particular care to inform the public through the popular press, with respect to how results from the eclipse expeditions would help to corroborate either of the two theories, viz., Newton's or Einstein's (cf. Sponsel, 2002).

Collins and Pinch (1998) have emphasized the importance of studying the dynamics of the construction of theories in science if we want to inform the citizens how scientists go through a process that involves arguments, controversies, and alternative interpretations. Stinner and Metz (2006) have advocated the use of

M. Niaz, *Critical Appraisal of Physical Science as a Human Enterprise: Dynamics of Scientific Progress*, Science and Technology Education Library 36,

thought experiments in introducing Einstein's theory of relativity. Furthermore, Brush (1999, p. 185) has cautioned that just because an experiment at present is seen to be confirming a theory, it cannot be assumed that it persuaded the scientific community when it was performed and hence the importance of studies that reconstruct historical episodes to show the complex relationship between observations, predictions, and theories.

The objective of this chapter is a critical appraisal of the various interpretations of the experimental evidence provided by the 1919 eclipse experiments for Einstein's General Theory of Relativity, and an attempt to provide closure to the controversy. This historical reconstruction is presented in the following sections:

1. Premise of the eclipse experiments
2. Experimental results reported by Dyson and Eddington
3. Salient aspects of the interpretation by Earman and Glymour (1980)
4. Salient aspects of the interpretation by Collins and Pinch (1993)
5. Salient aspects of the interpretation by Brush (1999)
6. Salient aspects of the interpretation by Hudson (2003)
7. Role of prediction (heuristic novelty) of bending of light and the eclipse experiments
8. Dyson and Eddington's contradictions
9. Role of presuppositions in understanding the eclipse experiments
10. Eddington's "implicit" presuppositions
11. Conclusion: Is closure possible?

Premise of the Eclipse Experiments

Stars can be seen close to the sun only during a solar eclipse. To determine the deflection of starlight, a scientist has to photograph a field of stars at a time when the sun is not between them and the Earth and photograph the same field during a total solar eclipse. The two sets of photographs would show the displacement of starlight due to the sun's gravitational field. The Sobral group had two instruments (an astrographic telescope and a 4-in. aperture telescope), and took 19 photographs with the astrographic telescope and eight with the 4-in. telescope. The Principe group had only one instrument (an astrographic telescope, similar to the one in Sobral) and took 16 photographs, most showing no star images and of these only two were usable. On the day of the eclipse, it was cloudy in both Sobral and Principe, which made observations difficult. The Sobral group took comparison photographs in July when the sun had moved away. The Principe group could not do the same due to transportation difficulties and had comparison photographs taken in Oxford. Estimate of the deflection to support Newton or Einstein is difficult to obtain as the scientist first waits for a total eclipse and then obtains comparison photographs (several months before or after) of the same region of the sky when the sun is absent. Weather conditions, temperature changes, and mechanical deformations of the telescopes can introduce serious experimental errors. Changes as small as a hundredth of a millimeter in the effective focal length of an astrographic telescope can produce significant effects in deflection

(Collins & Pinch, 1993; Dyson et al., 1920; Earman & Glymour, 1980). Thus, it is not surprising to note that experimental evidence from the eclipse expeditions "were not just a matter of looking through a telescope and seeing a displacement; they rested on a complex foundation of assumptions, calculations, and interpolations from two sets of photographs. And this is the case even if the photographs are clear and sharp – which they were not" (Collins & Pinch, 1993, p. 47). Given the complexity of the experiments, results obtained were open to various interpretations.

Experimental Results Reported by Dyson and Eddington

Based on the two expeditions (Sobral and Principe) the following sets of measurement of the deflection of starlight in the sun's gravitational field were estimated:

1. Sobral astrographic telescope mean	0.87″ (standard deviation = 0.48″)
2. Sobral 4-in. telescope mean	1.98″ (standard deviation = 0.178″)
3. Principe astrographic telescope mean	1.61″ (standard deviation = 0.444″)

Note: standard deviations reported here were calculated by Earman and Glymour (1980, p. 75) from the probable errors (a statistic no longer in use) reported in the original. These results were first reported at the joint meeting of the Royal Society and the Royal Astronomical Society on November 6, 1919. The meeting was chaired by J.J. Thomson, President of the Royal Society, who endorsed the results in the following terms:

> *It is difficult for the audience to weigh fully the meaning of the figures* that have been put before us, but the Astronomer Royal and Prof. Eddington have studied the material carefully, and they regard the evidence as decisively in favour of the larger value for the displacement. This is the most important result obtained in connection with the theory of gravitation since Newton's day. (Thomson, 1919, italics added)

In spite of the endorsement, the italicized part shows some reservation. Dyson, the Astronomer Royal, was categorical in supporting Einstein, especially based on results from Sobral 4-in. telescope: "[T]he much better plates [Sobral] gave for the displacement at the limb 1.98″, Einstein's predicted value being 1.75″.... After a careful study of the plates I am prepared to say that there can be no doubt that they confirm Einstein's prediction. A very definite result has been obtained that light is deflected in accordance with Einstein's law of gravitation" (Thomson, 1919, p. 391). A complete report of the findings was published by Dyson et al. (1920).

Salient Aspects of the Interpretation by Earman and Glymour (1980)

Earman and Glymour (1980) considered that given the importance of the issues at stake, the results should have been unequivocal, which they were not. Rather, results were conflicting and could be held to confirm Einstein's theory only if many of the measurements were ignored. After going through various critical aspects of the experiments they concluded:

> The natural conclusion from these results is that gravity definitely affects light, and that the gravitational deflection at the limb of the sun is somewhere between a little below 0.87″ and a little above 2.0″. If one kept the data from all three instruments [Sobral & Principe], the best estimate of the deflection would have to be somewhere between the Newtonian value and the Einstein value. If one kept only the results of the Sobral 4-inch instrument, the best estimate of the deflection would be 1.98″, significantly above even Einstein's value [approx. 1.3 standard deviations]. The conclusion that the Astronomer Royal announced to the extraordinary joint meeting at the astronomical and royal societies on November 6, 1919, was stronger: Einstein's prediction, Dyson announced, had been confirmed. (Earman & Glymour, 1980, p. 76)

Salient Aspects of the Interpretation by Collins and Pinch (1993)

According to Collins and Pinch (1993), one of the Sobral instruments (astrographic telescope) provided support for the Newtonian theory, while the other (4-in. telescope) leaned towards Einstein's prediction, and the two plates from Principe were the worst of all. Eddington on the other hand took the Sobral (4-in.) results as the main finding and used the two Principe plates as supporting evidence, while ignoring 19 plates from the Sobral astrographic on grounds of systematic errors. With this perspective, the authors asked: "Do the results come down on Einstein's side in an unambiguous way?" (p. 50) and responded in the negative and concluded:

> Eddington's observations, like many observations in science ... were very inexact and some of them conflicted with others. When he chose which observations to count as data, and which to count as "noise" ... Eddington had Einstein's prediction very much in mind. Therefore Eddington could only claim to have confirmed Einstein because he used Einstein's derivation in deciding what his observations really were, while Einstein's derivations only became accepted because Eddington's observation seemed to confirm them. (Collins & Pinch, 1993, p. 45)

Salient Aspects of the Interpretation by Brush (1999)

Brush (1999) points out that Einstein's theory made three testable predictions: (a) the advance of the perihelion of Mercury (Mercury's orbit); (b) the bending of light by a gravitational field; and (c) the gravitational redshift of spectral lines from a massive source. Brush argues that of these the first phenomenon (Mercury's orbit) had been known for several decades before the novel prediction of the bending of light. Furthermore, Newton's theory could not explain Mercury's orbit and hence its explanation by a rival theory (Einstein's) provides greater heuristic power than the bending of light by a gravitational field.

Salient Aspects of the Interpretation by Hudson (2003)

Hudson's (2003) main argument is that if Sobral astrographic data were discarded due to systematic errors based on the distortion of the coelostat due to the sun's heat, "why there was not a similar disruption with the Sobral 4-in. telescope along with its coelostat" (p. 125). This led Dyson and Eddington to violate the use-novelty condition as a methodological principle. Use-novelty condition is based on what Worrall (1985, 1989) and Giere (1983) have referred to as "heuristic novelty," viz., experimental evidence for a theory cannot have been used to create the theory itself. In the present case, Dyson and Eddington apparently rejected Sobral astrographic data as they did not provide support for Einstein's prediction of 1.74″ of starlight in the sun's gravitational field. In other words, Dyson and Eddington violated the use-novelty (heuristic novelty) condition as they used experimental evidence to construct a hypothesis, that is Sobral astrographic data were not reliable due to the coelostat distortion by the sun's heat. Mayo (1991, 1996, 2003), for example, supports the coelostat distortion hypothesis by justifying the unreliability of the Sobral astrographic results. Furthermore, Hudson (2003) reproduces astronomer Russell's (1920) criticism of Dyson's handling of the data, as photographs obtained with the Sobral 4-in. telescope also suffered from distortion. Finally, Hudson (2003) concluded:

> [C]irca 1920, the coelostat-distortion hypothesis (1) was not justified. Thus, constructing (1) on the basis of the observed data … is a violation of what I am terming *prima facie* use-novelty … on very general standards of evidence evaluation … it is clear that the published research reports of the eclipse experiments failed to make the case that the Sobral astrographic observations were unreliable. (Hudson, 2003, p. 118)

Role of Prediction (Heuristic Novelty) of Bending of Light and the Eclipse Experiments

Based on the Lakatosian framework, predictions (novel facts/heuristic novelty) play an important role in the development of scientific theories. A recent study has shown how D. Mendeleev's predictions could be interpreted as novel facts and thus played a crucial role in the development of the periodic table (Niaz et al., 2004). Lakatos (1970), in his Methodology of Scientific Research Programs, conceptualizes the role of predictions in the following terms:

(a) "The sophisticated falsificationist regards a scientific theory T as falsified if and only if another theory T′ has been proposed with the following characteristics: (1) T′ has excess empirical content over T: that is it predicts *novel* facts, that is, facts improbable in the light of, or even forbidden, by T" (Lakatos, 1970, p. 116, original emphasis; in a footnote on this page Lakatos adds that he uses "prediction" in a wide sense that includes "postdiction"). It is plausible to suggest that Einstein's contribution provided excess empirical content over that of Newton and furthermore made explicit prediction of the bending of light by 1.75″ in contrast to 0.87″ by Newton.

(b) "[A] series of theories is *theoretically progressive (or 'constitutes a theoretically progressive problem-shift')* if each new theory has some excess empirical content over its predecessor, that is, if it predicts some novel, hitherto unexpected fact" (p. 118, original emphasis). Einstein's contribution certainly complies with this requirement if we consider it as one of the many attempts to understand the bending of light in the sun's gravitational field (e.g., Newton, Lorentz, FitzGerald, and Abraham). Furthermore, according to Papineau (1979) the idea of progressiveness in Lakatos's methodology is closely related to predictions: "A programme is *empirically progressive* as well as theoretically progressive in so far as some of the new predictions are actually borne out" (p. 97, original emphasis).

(c) "The time-honoured empirical criterion for a satisfactory theory was agreement with the observed facts. Our empirical criterion for a series of theories is that it should produce new facts. *The idea of growth and the concept of empirical character are soldered into one*" (p. 119, original emphasis). Einstein's theory complies with this requirement partially as only the Sobral 4-in. telescope results provided fair amount of support for the theory. Mean deflection of the Sobral 4-in. telescope was 1.98″, which is significantly higher than Einstein's prediction of 1.75″ (cf. Earman & Glymour, 1980, p. 76).

(d) Lakatos raises a very pertinent issue: "*How are research programmes eliminated?*" (p. 155, original emphasis), and responds by suggesting "by a rival research programme which explains the previous success of its rival and supersedes it by a further display of *heuristic power*" (p. 155, original emphasis). In a footnote Lakatos explains: "I use '*heuristic power*' ... to characterize the power of a research programme to anticipate theoretically novel facts in its growth. I could of course use '*explanatory power*'" (p. 155, original emphasis). Einstein's theory clearly complies with this requirement as it predicts novel facts (heuristic novelty), provides greater explanatory power than Newton's theory, and significantly the novel fact was not used to construct the theory.

(e) It is important to note that Lakatos's idea of novel facts based on predictions has been the subject of considerable controversy in the literature (Frankel, 1979; Gardner, 1982; Lakatos & Zahar, 1978; Musgrave, 1974; Nunan, 1984; Thomason, 1992; Zahar, 1973). A detailed discussion goes beyond the scope of this study. Murphy (1989) has provided a synthesis and conceptualized the idea of novel facts in the following terms: "A fact that played no role in the formulation of the theory is one whose existence, relevance to the theory, or interpretability in light of the theory was first documented after that theory was first proposed" (p. 388). Once again, Einstein's theory complies with this requirement and he also looked forward to the eclipse experiments, contrary to the impression in some circles that he was indifferent to the results (cf. Hentschel, 1992). Furthermore, the prediction of novel facts within the Lakatosian framework comes quite close to heuristic novelty (Giere, 1983; Worrall, 1985, 1989), use-novelty, and *prima facie* use-novelty (Hudson, 2003; Mayo, 1991, 1996, 2003).

Dyson and Eddington's Contradictions

Previous sections of this study have shown how experimental evidence from the eclipse experiments has been far from unequivocal, which has led to various interpretations. To recapitulate: Of the three sources of experimental data, the Principe astrographic photographs were the worst, perhaps in part due to the cloudy weather. Data from the two Sobral telescopes were also affected by clouds and provided two different sets of measurements. The mean of the deflection from the Sobral 4-in. telescope was significantly higher than Einstein's prediction (1.98″ vs 1.75″), whereas the mean of the deflection from the Sobral astrographic telescope came quite close to the Newtonian prediction. Under these circumstances Dyson and Eddington adopted the following strategy: Deflection from Sobral 4-in. and Principe photographs was considered acceptable (both being close to the value predicted by Einstein), whereas the deflection from the Sobral astrographic telescope was rejected on grounds of systematic errors. This interpretation of the data leads to various contradictions:

(a) According to Dyson et al. (1920), on the day of the eclipse in Sobral (May 29, 1919), "the proportion of cloud was estimated at 9/10 at the time of first contact, when the sun was invisible" (p. 299). This decreased visibility but at the same time avoided the heating of the coelostat and thus could improve reliability of the results. In spite of this the authors stated that "there had been a serious change of focus [of the astrographic object glass]. ... This change of focus can only be attributed to unequal expansion of the mirror through the sun's heat" (p. 309). This, of course, raises the issue with respect to a similar effect of the sun's heat on the mirror of the other (4-in.) telescope at Sobral, whose results should also have been rejected. Nevertheless, as the Sobral 4-in. results were in the range of Einstein's prediction (the difference was, however, significant) they were accepted. Hudson (2003, p. 125) has also pointed out the same contradiction.

(b) Dyson et al. (1920) reported that due to bad weather, of the 16 photographs taken in Principe only two provided acceptable results and they refer to this as "the meagre material which the clouds permitted us to obtain" (p. 317). Based on check plates secured in Oxford and not in Principe, a displacement of 1.61″ (±0.30″) was estimated. Dyson et al. (1920) concluded the section (#37) on results from Principe as providing support for Einstein's theory.

(c) Dyson et al. (1920) soon after finishing section 37, started section 38 with the following statement: "It remains to consider the question of systematic error. The results obtained with a similar (to the one in Principe) instrument at Sobral (astrographic) are considered to be largely vitiated by systematic errors. What ground then have we – apart from the agreement with the far superior determination with the 4-inch lens at Sobral – for thinking that the present results [Principe] are more trustworthy? At first sight everything is in favor of the Sobral astrographic plates. There are 12 stars shown against 5, and the images though far from perfect are probably superior to the Principe images. ... Perhaps, therefore, the cloud was not so unkind to us after all" (pp. 328–329). This passage perhaps represents the crux of the whole issue. On the one hand, Dyson et al. (1920)

ignore the fact that on the day of the eclipse Sobral too was cloudy (i.e., equally kind) and that the difference between the 4-in. Sobral deflection (1.98″) and Einstein's prediction (1.75″) was significant.

(d) Here we reconstruct the reasoning of Dyson et al. (1920) in the last five pages as they try to summarize their results and convince the reader:

Step 1
"Nevertheless the accuracy seems sufficient [referring to Principe results] to give a fairly trustworthy confirmation of EINSTEIN'S theory, *and to render the half-deflection* [Newtonian value] *at least very improbable*" (p. 328, emphasis added). In spite of all difficulties associated with the Principe measurements (only 2 of the 16 photographs were usable), its results were still considered to be "trustworthy". Interestingly, the Newtonian value at this stage was not discarded but considered "very improbable".

Step 2
"Both [Sobral 4-inch and Principe] of these point to the full deflection 1.75″ of Einstein's generalised relativity theory, the Sobral results definitely, and the Principe results perhaps *with some uncertainty*. There remain the Sobral astrographic plates which gave the deflection 0.93″ discordant by an amount much beyond the limits of its accidental error" (p. 331, emphasis added). It is interesting to observe that the standard deviation of both Sobral astrographic and Principle measurements is the same (about 0.4″) and still results of the former are not even considered plausible but rejected. Furthermore, Sobral 4-inch results do not entirely agree with the Einstein prediction.

Step 3
"Thus the results of the expeditions to Sobral and Principe can *leave little doubt* that a deflection of light takes place in the neighbourhood of the sun and that it is of the amount demanded by Einstein's generalised theory of relativity, as attributable to the sun's gravitational field". (p. 332, emphasis added)

This constitutes an interesting piece of reasoning. In Step 1, Dyson et al. (1920) recognize that the Prinicpe results make the Newtonian half-deflection *at least very improbable*. Step 2 informs the reader that the Principe results are to be considered *with some uncertainty*. Step 3 provides the grand finale by accepting both the Sobral and Principe results, as they *leave little doubt* with respect to supporting Einstein's prediction. The sequence *at least very improbable* → *with some uncertainty* → *leave little doubt* starts with tautological reasoning and ends with a volte-face that leaves the reader perplexed.

Role of Presuppositions in Understanding the Eclipse Experiments

The controversy between R.A. Millikan and F. Ehrenhaft with respect to the oil drop experiment in order to determine the elementary electrical charge lasted for many years (1910–1925). In one of his major articles on the subject Millikan (1913) stated emphatically: "*It is to be remarked, too, that this is not a selected group of drops but represents all of the drops experimented upon during 60 consecutive days*" (p. 138, original italics). Philosopher-physicist G. Holton studied Millikan's handwritten notebooks at CALTECH, Pasadena, and reported that Millikan's (1913) study was based on 140 oil drops, whereas in the actual publication, data on 58 drops were reported. What happened to the remaining 82 drops, despite Millikan's assurance that data on all the drops studied were reported? Holton (1978) has explained succinctly how scientific theories are developed:

> It is generally true that prior to the absorption of research results into canonical knowledge, the selection of the relevant portion of a range of experience that is in principle infinite is guided by a hypothesis. That hypothesis in turn is stabilized chiefly by its success in handling that "relevant" portion and by the *thematic predisposition* which helps focus attention on it. (p. 211, emphasis added)

A recent study has attempted to provide closure to the controversial nature of the oil drop experiment by concluding that "if we try to understand Millikan's handling of data with no reference to his presuppositions, then some degree of 'misconduct' can be perceived" (Niaz, 2005b, p. 681, see Chapter 7). Presuppositions or "themata" thus play an important role in guiding scientists in the construction of theories. Support for this perspective is also provided by other philosophers of science, such as "negative heuristic of a research program" (Lakatos, 1970) and "guiding assumptions" (Laudan et al., 1988, also see Chapter 2).

With this background let us go back to the dilemma faced by Dyson and Eddington and consider the following scenario: Eddington and Dyson are not aware of Einstein's General Theory of Relativity and particularly of the prediction that starlight near the sun would bend. Under these circumstances experimental evidence from all three sources (Sobral and Principe) would have been extremely uncertain, equivocal, and difficult to interpret.

Let us now consider yet another scenario: W. Heisenberg and Einstein had a conversation during a walk in Berlin in April 1926 on the subject of electron orbits inside atoms and this is what Heisenberg (1969) reported (reproduced from Holton, 2000, p. 40):

> Heisenberg to Einstein: "Since it is acceptable to allow into a theory only directly observable magnitudes, I thought it more natural to restrict myself to these, bringing them in, as it were, as representatives of electron orbits"
>
> Einstein's response: "But you don't seriously believe that only observable magnitudes must go into a physical theory?"
>
> Heisenberg to Einstein: "I thought that it was exactly you who had made this thought the foundation of your relativity theory."
>
> Einstein's response: "Perhaps I used this sort of philosophy: but it is nevertheless nonsense. *Only the theory decides what one can observe*". (Emphasis added)
>
> Holton's comment: "Einstein, whose development away from positivistic instrumentalism to rational realism had escaped Heisenberg's notice, went on to explain at length how complicated any observation is in general, how it involves *assumptions about phenomena that in turn are based on theories*." (Emphasis added)

This conversation would have helped Eddington, who was otherwise fully aware of Einstein's General Theory of Relativity.

Eddington's "Implicit" Presuppositions

According to Warwick (1992), Eddington had early assumed the role of the unofficial champion of Einstein's theory. Thus, despite Eddington's considerable knowledge and personal stakes involved in defending Einstein's General Theory of Relativity, at no place in Dyson et al. (1920) (the authoritative publication of the eclipse experimental results) do they explicitly refer to their presuppositions. They do, however, cast Einstein's theory in a positive light and throughout the article mention it 21 times, in comparison to seven times for Newton's theory. In this respect they followed the same methodological strategy as that of Mendeleev (periodic table, cf. Niaz et al., 2004, Chapter 5) and Millikan (oil drop experiment, cf. Niaz, 2005b, Chapter 7). Nevertheless, there is ample evidence to show that they do refer to "implicit" presuppositions. For example, on the one hand, Dyson et al. (1920) recognize the problematic nature of the Principe measurements, and still insist that these "render the half-deflection [Newtonian value] at least very improbable" (p. 328). Furthermore, throughout the article, it seems that any deflection higher than the Newtonian value automatically constitutes evidence for Einstein's theory. Results from the Sobral 4-in. telescope were significantly higher than the Einstein prediction, and still Dyson et al. (1920) simply ignore this fact. It is plausible to suggest that in the absence of presuppositions, none of the two theories (Newton or Einstein) could have been corroborated, and the most reasonable conclusion would have been that starlight is affected by the sun's gravitational field.

Based on their "implicit" presuppositions, Dyson et al. (1920) summarized their results in the following terms: "The results of the observations here described appear to point quite definitely to the third alternative [full deflection], and confirm Einstein's generalised theory of relativity" (p. 292). Similar statements are made in two other places in the article (pp. 328, 331), which shows the degree to which Eddington's experimental data were "implicitly" based on the premise (presupposition) that Einstein's General Theory of Relativity was correct. This conceptualization of confirmation of a theory by empirical evidence is problematic. Eddington's data could have confirmed Einstein's prediction, but that does not mean that Einstein's theory was also confirmed. Due to the underdetermination of scientific theories by experimental data (Duhem–Quine thesis, cf. Quine, 1953), science does not advance by just having the data. Philosopher-physicist J. Cushing (1989) has explained this in cogent terms: "One cannot simply amass a 'Baconian' heap of facts (or of theses) and then hope that truth or a theory will thereby emerge" (pp. 19–20). Lakatos (1999) has stated the same point much more forcefully: "Inductivism claims that a proposition is scientific if provable from facts; what we shall now set out to do is to show that no proposition whatsoever can be proven from facts. And certainly not any scientific proposition" (p. 36).

Within a historical perspective it is important to note how P. Duhem (1861–1916) has emphasized that experiments in physics involve not simply the observation of a phenomenon but also its theoretical interpretation:

> The same practical fact may correspond to an infinity of logically incompatible theoretical facts; the same group of concrete facts may be made to correspond in general not with a single symbolic judgment but with an infinity of judgments different from one another and logically in contradiction with one another. (Duhem, 1914, p. 152)

Without subscribing to Duhem's philosophy of science entirely, it appears that in the present context Duhem's perspective is relevant, as he represents the same scientific community as Eddington, and thus the two being contemporaries shared similar scientific practices (cf. Brush, 1999, p. 185, for the need to evaluate experiments according to the perspective of the scientific community).

Conclusion: Is Closure Possible?

It is plausible to suggest that Eddington's (Dyson et al., 1920) methodology approximated the traditional scientific method, which did not allow him to discard discrepant data. Dyson and especially Eddington were fully aware as to where the theory (Einstein's) was leading them. Nevertheless, given the inductivist/positivist milieu of their scientific community they could only be guided by their "implicit" presuppositions. Following any other alternative course involved the risk of having all their experimental data being declared ambiguous and ultimately invalid. This weighed more in this case as eclipse experiments are not easy to replicate (contrast this with Millikan's oil drop experiment). Millikan in his publications espoused the scientific method but in his private handwritten notebooks (Holton, 1978) was fully aware of the dilemma faced and was forced to select data to uphold his presuppositions (Niaz, 2005b). A closure is possible if we grant Eddington the possibility of being guided by his "implicit" presuppositions, which were based on, and further reinforced due to, the following:

(a) Einstein's theory provided excess empirical content over that of Newton's (cf. Lakatos, 1970, p. 116).
(b) Einstein's theory provided a progressive problemshift (cf. Lakatos, 1979, p. 118) based on bending of light by 1.98″ from data obtained by the Sobral telescope). Eddington's dilemma was that none of the three telescopes provided a value close to that predicted by Einstein, i.e., 1.75″. In this sense, his "implicit" presuppositions were all the more crucial and provided grounds for discarding data from the Sobral astrographic telescope.
(c) Prediction of a novel fact (heuristic novelty) by Einstein's theory played no role in the formulation of the theory itself (cf. Lakatos, 1970, p. 155; Murphy, 1989, p. 388). Dyson et al. (1920) allude to this prediction in the following terms: "They [data from deflection of light] agree with Einstein's *predicted* value 1.75″" (p. 327, italics added). This statement is crucial in understanding Eddington's dilemma, as Brush (1989) had stated: "It is remarkable that the authors [Dyson et al., 1920] avoid using any form of the word 'predict' in the text of this paper" (footnote 30, p. 1128).

Finally, a recent attempt to study the structure of scientific controversies shows that the closure point cannot be fixed by logical or methodological rules and finally recognizes the role of "presuppositions": "What science says the world is like at a certain time is affected by the human ideas, choices, expectations, prejudices, beliefs, and assumptions holding at that time" (Machamer et al., 2000, p. 6). Indeed, Eddington's handling of the eclipse experiments and interpretation of the data was very much in tune with the scientific and philosophical milieu of the early twentieth century.

Chapter 10
Lewis's Covalent Bond: From Transfer of Electrons to Sharing of Electrons

Introduction

Lewis (1916) is generally considered to have presented the first satisfactory model of the covalent (shared pair) bond based on the cubic atom in 1916. It is important to note that the genesis of the cubic atom can be traced to an unpublished memorandum written by Lewis in 1902 and recounted by him in the following terms:

> In the year 1902 (while I was attempting to explain to an elementary class in chemistry some of the ideas involved in the periodic law) becoming interested in the new theory of the electron (Thomson's discovery of the electron in 1897), and combining this idea with those which are implied in the periodic classification, I formed *an idea of the inner structure of the atom* (model of the cubic atom) which, although it contained crudities, I have ever since regarded as representing essentially the arrangement of the electrons in the atom. (Lewis, 1923, pp. 29–30, emphasis added)

The objective of this chapter is to understand how the postulation of the covalent bond was considered to be controversial and had to compete with the well-established ionic bond. This historical reconstruction is presented in the following sections:

1. Lewis's 1902 model of the cubic atom
2. Lewis's model of the covalent bond
3. Historical antecedents of the covalent bond
4. Origin of the covalent bond: a "Baconian inductive ascent"?
5. Educational implications of the historical reconstruction of the covalent bond

Lewis's 1902 Model of the Cubic Atom

Lewis (1916, p. 768) reproduced the following postulates of his 1902 theory of the cubical atom at length in his 1916 article:

1. In every atom there is an essential *kernel*, which remains unaltered in all ordinary chemical changes and which possesses an excess of positive charges corresponding

M. Niaz, *Critical Appraisal of Physical Science as a Human Enterprise: Dynamics of Scientific Progress*, Science and Technology Education Library 36,

in number to the ordinal number of the group in the periodic table to which the element belongs.

2. The atom is composed of the kernel and an *outer atom* or *shell*, which in the case of the neutral atom contains negative electrons equal in number to the excess of positive charges of the kernel, but the number of electrons in the shell may vary during change between 0 and 18.
3. The atom tends to hold an even number of electrons in the shell, and especially to hold eight electrons, which are normally arranged symmetrically at the eight corners of a cube.
4. Two atomic shells are mutually interpenetrable.
5. Electrons may ordinarily pass with readiness from one position in the outer shell to another. Nevertheless they are held in position by more or less rigid constraints, and these positions and the magnitude of the constraints are determined by the nature of the atom and of such other atoms as are combined with it.
6. Electric forces between particles which are very close together do not obey the simple law of inverse squares, which holds at greater distances.

After having presented the six postulates, Lewis (1916) elaborated by providing further information. For example:

> The first postulate deals with the two parts of the atom which correspond roughly with the inner and outer rings of the Thomson atom. The kernel being that part of the atom which is unaltered by ordinary chemical change. (p. 768)

Postulate 2 was illustrated by indicating how chlorine has eight electrons in the outer shell, while forming chlorides. Postulate 3 was the most striking and at the same time controversial feature of Lewis's theory, which led to the formulation of the "rule of eight" or the "octet rule." The rule of eight as proposed by Lewis differed from the "law of octaves" proposed by Newlands in 1865, according to which the elements when listed in the order of increasing atomic weights in the periodic table, the eighth element would be similar to the first. Lewis postulated that the eight electrons of an octet formed the eight corners of a cube, as this provided "the most stable condition for the atomic shell" (p. 774). Thus, the single bond was conceived of as two cubic atoms with a shared edge (pair of electrons, see Fig. 10.1) and the double bond as two cubes with a common face (see Fig. 10.2). Postulate 4 facilitated further the idea of sharing of electrons: "Thus an electron may form a part of the shell of two different atoms and cannot be said to belong to either one exclusively" (p. 772). Interestingly, in elaborating on Postulate 5, Lewis disagreed with Bohr's (1913a) model of the atom in the following terms:

> It seems to me far simpler to assume that an electron may be held in the atom in stable equilibrium in a series of different positions, each of which having definite constraints, corresponds to a definite frequency of the electron. (p. 773)

Postulate 6 reinforced a point of view that coincided with new developments in the structure of the atom: "[A] study of the mathematical theory of the electron leads, I believe, irresistibly to the conclusion that Coulomb's law of inverse squares must fail at small distances" (p. 773).

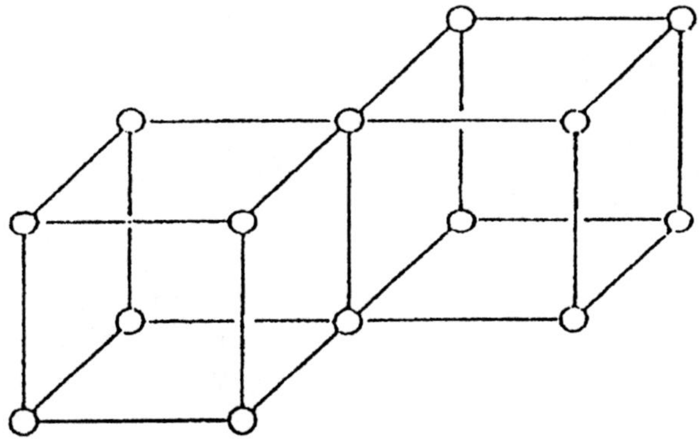

Fig. 1 Two cubes with a shared edge (single bond)

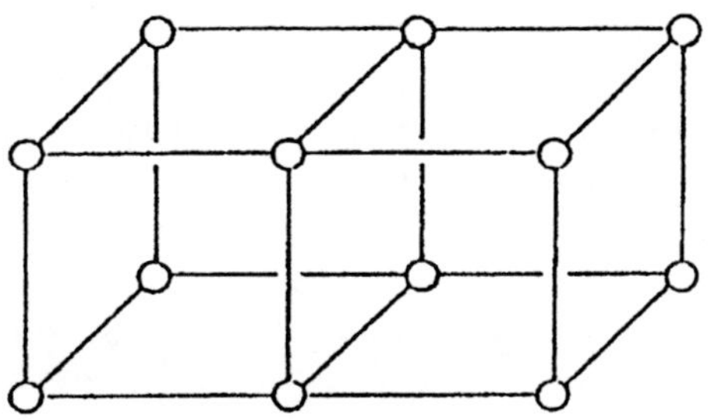

Fig. 2 Two cubes with a common face (double bond)

Lewis's Model of the Covalent Bond

In this section evidence is provided to show that Lewis's theory of sharing electrons (covalent bond) had to compete with a rival theory, viz., transfer of electrons (ionic bond). From a philosophy of science perspective the rivalry between competing theories (paradigms/research programs) is an integral part of scientific progress. According to Lakatos (1970):

> [R]esearch programmes have achieved complete monopoly only rarely and then only for relatively short periods. ... The history of science has been and should be a history of competing research programmes. (p. 155)

According to Kohler (1971), who has presented a detailed account of the origin of Lewis's ideas:

> When it was first proposed, Lewis's theory was completely out of tune with established belief. For nearly 20 years it had been almost universally believed that all bonds were formed by the complete transfer of *one* electron from one atom to another. The paradigm was the ionic bond of $Na^+ Cl^-$, and even the bonds in compounds such as methane or hydrogen were believed to be polar, despite their lack of polar properties. From the standpoint of the polar theory the idea that two negative electrons could attract each other or that two atoms could share electrons was absurd. (p. 344)

Rodebush (1928), a chemist reviewing the origin of the covalent bond in the late 1920s, shared the same concern:

> Since according to Coulomb's law two electrons should exert a repulsion for each other, the pairing of electrons seems at first glance to be a bizarre idea. In order to account for the peculiar behavior Lewis assumed the existence of a magnetic attraction between the electrons. (pp. 513–514)

Lewis (1916) further clarified his attempt at building a theory of the atom: "In my original theory [1902] I considered the elements in the periodic table thus built up, as if block by block, forming concentric cubes" (p. 769). Later in the article, Lewis recognizes that the cubic structure cannot represent the triple bond and suggests its replacement by the tetrahedral atom (p. 780).

At this stage it is important to note that Thomson's (1897) discovery of the electron in 1897 and later publications (Thomson, 1907) provided powerful arguments for the polar theory of the ionic bond. According to Thomson (1907): "For each valency bond established between two atoms the transference of one corpuscle from the one atom to the other has taken place" (p. 138). Although Thomson accepted that overlapping of corpuscles could produce a nonpolar bond in theory, he believed that in reality all bonds were polar bonds (p. 131). One of the best known theories of electrovalence representing the polar orthodoxy was proposed by Abegg (1904). According to this theory the only kind of chemical affinity was electrostatic attraction, and all bonds, even in nonpolar symmetrical molecules, were electron transfer bonds.

Historical Antecedents of the Covalent Bond

According to Kohler (1971), although sharing of electrons to form covalent bonds seemed shocking at first, few chemists have shown interest in the origin of the shared pair bond. In the previous sections it has been shown that the cubic atom was important in the development of Lewis's theory of the shared pair covalent bond. Lewis's cubic atom was first conceived as a teaching device to illustrate the octet rule and can be considered as "speculative." Modern philosophy of science has emphasized the importance of speculations as a part of the scientific process (Kuhn, 1970; Lakatos, 1970). For example, Bohr's use of the quantum postulate in his theory is considered to be speculative. Interestingly, Lewis (1919) in a letter to Robert Millikan complained:

> I could not find a soul sufficiently interested to hear the theory [1902 memorandum]. There was a great deal of research work being done at the university [Harvard], but, as I see it now, the spirit of research was dead.

First dissenting voices against the polar orthodoxy were raised in 1913 by Bray and Branch (1913), colleagues of Lewis at MIT. Bray and Branch (1913) objected that the polar theory had been extended far beyond its proper limits and suggested:

> [T]here are two distinct types of union between atoms: polar, in which an electron has passed from one atom to the other, and non-polar, in which there is no motion of an electron. (p. 1443)

According to Kohler (1971): "By 1913 or so the polar theory completely dominated chemistry, and it did until it was replaced by Lewis's theory in the early 1920s" (p. 355). After the lead of Bray and Branch, Lewis himself published two papers (1913, 1916).

Arsem (1914), a student of Bray at MIT, was the next to publish with the following introductory comments: "I have … ventured to present my views at this time with a feeling that they are in harmony with the present trend of scientific speculation" (p. 1656). Arsem (1914) presented his critique in the following terms:

> There is a difficulty in accounting, on the basis of Thomson's theory [1907], for the existence of a hydrogen molecule made up of two positive atoms, or of a chlorine molecule with two negative atoms, and it is hard to see why two neutral atoms or two atoms having equal valency of the same sign should combine to form a stable molecule. (p. 1658)

Arsem's model of the hydrogen molecule was based on a single electron that oscillated between the two atoms. The oscillating electron was an intrinsic part of both atoms at the same time. Although Arsem's "half-electron bond" was considered to be flawed, it stopped just one step short of the two-electron-shared pair bond. Later, Thomson (1914) himself changed and accepted that all bonds were not polar bonds after all. Thomson conceived the nonpolar bond as a tube of force stretching from an electron in one atom to the nucleus of a second.

Parson (1915) was the next to contribute towards the development of the covalent bond. Following the French physicist Langevin (1872–1946), he conceived the idea that the force responsible for chemical bonding was not electrical but magnetic. Langevin had proposed in 1905 that the electron was a zero-resistance electric circuit, i.e., an electromagnet (magnetons). Parson's atom consisted of a sphere of positive charge (Thomson, 1907) in which the magnetons were arranged not in concentric rings but at the corners of cubic octets, reminiscent of Lewis's 1902 cubical atom. Interestingly, Parson finished his manuscript while spending a year at Berkeley with Lewis. Apparently, Parson had only two interviews with Lewis, and in one of the interviews Lewis drew a cubic atom and remarked: "I once had the idea of a cube corresponding to the octave law" (reproduced in Kohler, 1971, p. 370).

Kohler's reconstruction of Lewis's theory of the covalent bond presented above contrasts with the interpretation of Rodebush (1928), a chemist who was reviewing the literature on the development of the covalent bond. Rodebush considers Lewis's shared pair covalent bond to have been induced from two empirical facts: (i) Moseley's demonstration that helium, an inert gas, has a pair of electrons; and (ii) the fact that with a few exceptions the number of electrons in all compounds is even. According to Kohler (1971): "Rodebush's empiricistic view reveals what chemists at the time thought their science ought to be like" (p. 371). This shows how Lewis's work was incorrectly interpreted even in the late 1920s.

Origin of the Covalent Bond: A "Baconian Inductive Ascent"?

At this stage it is instructive to study how philosophers of science have interpreted Bohr's model of the atom. According to Lakatos (1970):

> Bohr's problem was not to explain Balmer and Paschen series, but to explain the paradoxical stability of the Rutherford atom. Moreover, Bohr had not even heard of these formulae before he wrote the first version of his paper. (p. 147)

This version of the events has been corroborated by an extremely careful and detailed study by Heilbron and Kuhn (1969). Interestingly, most textbooks consider Bohr's major contribution to be the explanation of the Balmer and Paschen series of the hydrogen line spectrum. On the contrary, philosophers of science consider Bohr's explanation of the paradoxical stability of the Rutherford model of the atom as his major contribution (Heilbron & Kuhn, 1969; Lakatos, 1970).

In a study based on 23 general chemistry textbooks, Niaz (1998) found that very few textbooks considered the explanation of the paradoxical stability of the Rutherford model of the atom as important and none of the textbooks interpreted the quantization of the Rutherford model within a historical perspective.

Lakatos (1970, p. 147) goes on to show the importance of this event in the history of science: "Since the Balmer and the Paschen series were known before 1913 [year of Bohr's first publication], some historians present the story as an example of a Baconian 'inductive ascent': (1) the chaos of spectrum lines, (2) an empirical law (Balmer), (3) the theoretical explanation (Bohr)."

A major premise of historians who follow the Baconian inductive ascent is that scientific theories and laws are primarily driven by experimental observations. Furthermore, such empiricist interpretations consider scientific progress to be dichotomous, viz., experimental observations lead to scientific laws, which later facilitate the elaboration of explanatory theories.

A major thesis of this chapter is that the conceptualization of the covalent bond by chemists and textbooks approximates quite closely to a Baconian inductive ascent, according to the following stages:

1. The finding that diatomic molecules such as H_2 and the hundreds of compounds found by organic chemists in the late nineteenth century could not be understood by the ionic bond (this corresponds to the chaos of spectrum lines before Balmer's law).
2. Postulation of the nonpolar shared pair covalent bond by Lewis as an inductive law or generalization (this corresponds to Balmer's empirical law for the hydrogen line spectrum).
3. Theoretical explanation offered by quantum theory (Pauli exclusion principle) as to how two electrons (in spite of the repulsion) can occupy the same space (this corresponds to Bohr's explanation of the hydrogen line spectrum). This aspect is corroborated by the following interpretation of the events by Rodebush (1928):

> It will be recognized by the chemist however that Pauli's rule [exclusion principle] is only a short hand way of saying what Lewis has assumed for many years as the basis of his magnetochemical theory … of valence. If the electrons are paired in the atom magnetically,

> it is easy to see how two unpaired electrons in different atoms may be coupled magnetically and form the nonpolar bond. (p. 516)

The crux of the issue is that the inductivist interpretation construes Pauli's exclusion principle as the theoretical explanation and ignores the fact that Lewis's cubic atom was crucial for his later explanation of the sharing of electrons. Thus, scientific progress is characterized by a series of theories or models (plausible explanations) which vary in the degree to which they explain the experimental findings. In other words, science does not necessarily progress from experimental findings to scientific laws to theoretical explanations. According to Lakatos (1970, p. 129) the conflict is not between theories and laws, but rather between an interpretative and an explanatory theory.

Educational Implications of the Historical Reconstruction of the Covalent Bond

Based on the previous sections it would be interesting to evaluate general chemistry textbooks on the following heuristic principles (criteria).

Lewis's Cubic Atom as a Theoretical Device for Understanding the Sharing of Electrons

Niaz (2001b) has reported that of the 27 general chemistry textbooks (published in the USA) analyzed, only three described satisfactorily the role played by Lewis's cubic atom and following is an example:

> Lewis assumed that the number of electrons in the outermost cube on an atom was equal to the number of electrons lost when the atom formed positive ions ... he assumed that each neutral atom had one more electron in the outermost cube than the atom immediately preceding it in the periodic table. Finally, he assumed it took eight electrons – an octet – to complete a cube. Once an atom had an octet of electrons in its outermost cube, this cube became part of the core, or kernel, of electrons about which the next cube was built. (Bodner & Pardue, 1989, p. 273)

With this introduction, authors explain the formation of the covalent bond:

> By 1916, Lewis had realized that there was another way atoms could combine to achieve an octet of valence electrons – they could share electrons. Two fluorine atoms, for example, could share a pair of electrons and thereby form a stable F_2 molecule in which each atom had an octet of valence electrons. (p. 274)

Authors provide pictures of the individual cubes of fluorine coming together to share an edge and thus form the covalent bond in which the eight electrons are oriented toward the corners of a cube. Furthermore, the authors reproduce Lewis's 1902 memo with handwritten drawings of the cubic atom that were included by Lewis (1923) in his now famous book on valence. Interestingly, another textbook (Chang, 1981, p. 156) also reproduced the handwritten memo (about half a page) with no reference to the significance of the cubic atom, or as to what Lewis was trying to do.

Sharing of Electrons (Covalent Bond) Had to Compete with the Transfer of Electrons (Ionic Bond)

One of the textbooks mentions but none describes satisfactorily that Lewis's idea of sharing electrons had to compete with the transfer of electrons (Niaz, 2001b). A number of textbooks start with the presentation of the covalent bond in terms that can be useful. Nevertheless, these authors do not interpret the origin of the covalent bond as a rival research program and following is an example:

> The vast majority of chemical substances do not have the characteristics of ionic materials; water, gasoline, banana peelings, hair, antifreeze, and plastic bags. ... For the very large class of substances that do not behave like ionic substances, we need a different model for the bonding between atoms. Lewis reasoned that an atom might acquire a noble-gas electron configuration by sharing electrons with other atoms. A chemical bond formed by sharing a pair of electrons is called a *covalent bond*. (Brown & LeMay, 1988, p. 233)

A brief reference to the historical details can facilitate conceptual understanding and make students aware of the difficulties involved and how scientists (like students) resist change.

Covalent Bond: Inductive Generalization/Derived from the Cubical Atom

The objective of this criterion was to evaluate if the textbooks follow one of the following interpretations with respect to the origin of the covalent bond: (a) inductivist – Lewis's covalent bond was an inductive generalization based on the fact that stability of the noble gases or formation of the hydrogen molecule leads to a lowering of the energy, or that Helium, an inert gas, has a pair of electrons; (b) Lakatosian – Lewis's covalent bond was not induced from experimental evidence but derived from the cubic atom. Niaz (2001b) has reported that 23 textbooks considered the origin of the covalent bond to be an inductive generalization and following is an example:

> Because all noble gases (except He) have eight valence electrons, many atoms undergoing reactions also end up with eight electrons. This observation had led to what is known as the *octet rule*. (Brown & LeMay, 1988, p. 225)

Such presentations are quite representative of most textbooks and show clearly that the octet rule is sustained by empirical evidence. The alternative interpretation (Lakatosian) would have emphasized the cubic atom – a hypothetical entity. Only two textbooks (Bodner & Pardue, 1989; Mahan & Myers, 1987) presented the

Lakatosian interpretation. These textbooks trace the origin of the stability of the covalent bond to the cubic atom and go to considerable length to show that Lewis's ideas developed slowly, were based on conjectures, and were tentative. At first sight the difference between the two approaches may seem trivial. Nevertheless, the inductivist approach considers all scientific findings to be driven by experiment, which comes quite close to a "Baconian inductive ascent" (cf. Niaz, 1999b). It is important to note that it was Lewis's model of the cubic atom and not the octet rule that provided the first tentative (conjectural) explanation of the covalent bond. A better explanation ("progressive problemshift," Lakatos, 1970) is provided by Pauli's exclusion principle (Pauli, 1925), which is the subject of the next criterion.

Pauli's Exclusion Principle as an Explanation of the Sharing of Electrons in Covalent Bonds

Niaz (2001b) has reported that five textbooks mentioned and eight described satisfactorily Pauli's principle as an explanation of the sharing of electrons (with opposite spin) in covalent bonds and following is an example:

> When two hydrogen atoms form a covalent bond, the atomic orbitals overlap in such a way that the electron clouds reinforce each other in the region between the nuclei, and there is an increased probability of finding an electron in this region. According to Pauli's exclusion principle, the two electrons of the bond must have opposite spins. The strength of the covalent bond comes from the attraction of the positively charged nuclei for the negative cloud of the bond. (Mortimer, 1983, p. 135)

A historical reconstruction shows that the Lewis's cubic atom was the first attempt to explain the stability of the covalent bond and later the quantum theory provided further support when Pauli introduced his exclusion principle. The transition from Lewis's cubic atom → Pauli's exclusion principle → what next provides an illustration of how scientific knowledge is tentative.

Development of the Covalent Bond as a "Baconian Inductive Ascent"

Presentations of some of the textbooks came quite close to what could be considered as a Baconian inductive ascent, as the following two examples demonstrate (Niaz, 2001b):

> Most of the reasons for matter bonding to matter ... can be summarized by two concise notions: (i) Unlike charges attract; (ii) Electrons tend to exist in pairs. Couple these two ideas (one empirical, one theoretical) with the proximity requirement that only the outer electrons of the atoms (the *valence electrons*) interact, and you have the basic concepts that explain how atoms in over 10 million compounds bond to each other. (Joesten et al., 1991, p. 109, original emphasis)

> In observing millions of stable compounds, chemists have learned that *in almost all stable chemical compounds of the representative elements, all of the atoms have achieved a noble gas electron configuration.* The importance of this observation cannot be overstated. It forms the basis of our fundamental ideas about why and how atoms bond to each other. (Zumdahl, 1990, p. 373, original emphasis)

These are good examples of the first two steps of the Baconian inductive ascent, namely, chaos of experimental observations led to the postulation of the covalent bond. Nevertheless, these two textbooks did not go on to the third step – explanation of the covalent bond by Pauli's exclusion principle. Interestingly, of the eight textbooks that did consider Pauli's exclusion principle as a satisfactory (S) explanation of the covalent bond, only two referred to Lewis's cubic atom. This shows that lack of a history and philosophy of science perspective perhaps leads the textbooks to recognize the role of Pauli's exclusion principle and yet not even mention Lewis's cubic atom.

Blanco and Niaz (1997) found that many chemistry teachers and students consider progress in science to be characterized by a "Baconian inductive ascent," that is, experimental findings → scientific laws → theoretical explanations. Furthermore, Blanco and Niaz (1997) concluded that this finding can provide "[a] plausible blueprint for alternatives to the traditional textbook treatment of progress in science" (p. 228). The blueprint for an alternative approach in the present case would be a textbook presentation emphasizing the origin of the covalent bond as a product of conflicting or rival theories (models) for the explanation of bond formation. This shows that appropriate historical reconstructions can benefit students both by providing them with models for alternative/rival approaches and by instilling in them a deeper conceptual understanding of the topic (cf. Chiappetta et al., 1991).

Chapter 11
Quantum Mechanics: From Bohr to Bohm

Introduction

Quantum numbers and electron configurations of chemical elements form an important part of the physical science curriculum and textbooks devote considerable amount of space to these topics. These topics are closely related to students' understanding of quantum mechanics and various studies have reported students' difficulties in grasping the fundamental issues involved (Shiland, 1995, 1997; Tsaparlis, 1997, 2001; Johnston et al., 1998; Dobson et al., 2000; Hadzidaki et al., 2000; Ireson, 2000; Michelini et al., 2000; Pospiech, 2000; Ardac, 2002; Wittmann et al., 2002; Kalkanis et al., 2003; Taber, 2005; Niaz, 2008; Niaz & Fernández, 2008). Interestingly, physicists have also recognized the difficulties involved in understanding quantum mechanics (Feynman, 1985; Styer, 2000; Laloë, 2001). Feynman (1965) was quite categorical: "I can safely say that nobody understands quantum mechanics" (p. 129). On the other hand, philosophers of science have argued that quantum mechanics is particularly difficult to understand, due to the controversial nature of the different interpretations (e.g., Bohr's Copenhagen "indeterminacy" and Bohm's "hidden variables"). According to physicist-philosopher Abner Shimony (1985), "I must confess that after 25 years of attentive – and even reverent – reading of Bohr, I have not found a consistent and comprehensive framework for the [Copenhagen] interpretation of quantum mechanics" (p. 109). In contrast, in a recent critical review a physicist has conceded that:

> At the turn of the century, it is probably fair to say that we are no longer sure that the Copenhagen interpretation is the only possible consistent attitude for physicists.... Alternative points of view are considered as perfectly consistent: theories including additional variables (or "hidden variables"). (Laloë, 2001, p. 656)

Cushing (1991) has expressed the crux of the issue in cogent terms:

> The question is whether we are capable of truly *understanding* (or comprehending) quantum phenomena, as opposed to simply *accepting* the formalism and certain irreducible quantum correlations. The central issue is that of understanding versus merely redefining terms to paper over our ignorance. (p. 337, original italics)

M. Niaz, *Critical Appraisal of Physical Science as a Human Enterprise: Dynamics of Scientific Progress*, Science and Technology Education Library 36,

In order to facilitate understanding, Cushing (1991, pp. 337–338) has suggested that in the physical sciences scientific theories function at the following three levels: (a) *empirical adequacy* – consists of essentially in "getting the numbers right," in the sense of having a formula or an algorithm that is capable of reproducing observed data; (b) *explanation* – this is provided by a successful formalism with a set of equations and rules for its application; and (c) *understanding* – this is possible once we have an interpretation of the formalism that allows us to comprehend and to know the character of the phenomena and of the explanation offered. In order to illustrate these three levels, Cushing (1991) provides three examples drawn from laws of motion (Kepler and Newton), gas laws (Boyle and Maxwell), and quantum mechanics. In the case of gases our understanding was facilitated by the following levels of scientific development: Boyle's law → formalism of statistical mechanics → kinetic theory. Boyle's law is simply a phenomenological law summarizing empirical data, statistical mechanics gives us no understanding of the physical mechanism involved in gases. It is the kinetic theory that provides us an understanding of the behavior of gases under different experimental conditions. In the case of quantum mechanics, empirical adequacy would refer to the early work by Bohr and colleagues and Einstein's response based on a thought experiment, EPR (Einstein et al., 1935). Fair amount of literature on this phase of quantum mechanics is available (cf. Brush, 1980; Fine, 1996b). The next level would be the development of the formalism of quantum mechanics (as presented in most textbooks), and this would constitute the explanatory phase. In this sequence, Bohm–Vigier interpretation would perhaps provide insight at precisely the level of understanding (Bohm & Vigier, 1954; Vigier, 1982). Similarly, Styer (2000, p. xiii) has emphasized the need to delve deeply into quantum mechanics as a set of physical ideas rather than as an elaborate and somewhat mystical algorithm for solving problems in atomic physics. The objective of this chapter is to facilitate alternative interpretations of quantum mechanics, and this historical reconstruction is presented in the following sections:

1. Origin of the quantum hypothesis
2. A brief introduction to the life and career of Bohm
3. Origin and development of Bohm's interpretation of quantum mechanics
4. Comparison of Bohmian and Bohr's Copenhagen interpretations of quantum mechanics
5. Educational implications of the historical reconstruction

Origin of the Quantum Hypothesis

It is well known that Thomas Kuhn directed the Project, "Sources for History of Quantum Physics," a valuable archive now available at various institutions around the world. Besides a major publication (Heilbron & Kuhn, 1969), the only other scholarly work that was based on the sources of this "Project" was about *Black-body theory* (Kuhn, 1978). However, it is generally neglected (perhaps due to a major interest in paradigms) that in this later work, Kuhn raised a provocative question:

Who first proposed the quantum hypothesis? More recently, Brush (2000) has brought attention to this facet of Kuhn's work and its implications for physics textbooks. Kuhn (1978) stated quite categorically that:

> [T]he arguments in Planck's first quantum papers [Planck, 1900] did not, as I now read them, seem to place any restrictions on the energy of the hypothetical resonators that their author had introduced to equilibrate the distribution of energy in the black-body radiation field. Planck's resonators, I concluded, absorbed and emitted energy continuously at a rate governed precisely by Maxwell's equations. His theory was still classical. (p. viii)

Kuhn concluded that it was Paul Ehrenfest and Albert Einstein who first recognized that the black-body law could not be derived without restricting resonator energy to integral multiples of $h\nu$. In other words, Planck in 1900 simply introduced an approximate mathematical quantization for convenience in doing the calculations. On the other hand, the physical significance of the quantum hypothesis was first explained by Einstein. Despite skepticism on the part of some historians with respect to Kuhn's controversial thesis, Brush (2000) has shown that a historical reconstruction provides a clear confirmation of Kuhn's thesis. A recent appraisal of quantum mechanics has also endorsed Kuhn's thesis:

> Whereas Planck, ... became a revolutionary against his will, Einstein recognized the revolutionary implications of the quantum hypothesis much more clearly and willingly acted as a prophet of the quantum revolution. There is much truth in the claim that quantum theory started in earnest only in 1905, with Einstein's works. (Kragh, 1999, p. 66)

In the context of Cushing's (1991) scheme, it is plausible to suggest that Planck's contribution in 1900 looks more like at the level of empirical adequacy, whereas Einstein's contribution provided some degree of explanation/understanding, and Bohmian mechanics would go further to facilitate understanding.

A Brief Introduction to the Life and Career of David Bohm

David Bohm (1917–1992) was born in Wilkes-Barre, Pennsylvania, a center of coal-mining and manufacturing industries and attended Pennsylvania State University.[1] After doing graduate work at the California Institute of Technology, he moved to University of California, Berkeley, where he studied theoretical physics with J. Robert Oppenheimer, receiving his Ph.D. in 1943. While at the Berkeley Radiation Laboratory, he worked on problems of uranium separation. Bohm's interest in philosophy and Marxism led him to join the Communist Party from 1942 to 1943. He was also active in the Federation of Architects, Engineers, Chemists, and Technicians (FAECT), a left-wing union, affiliated with the Congress of Industrial Organizations (CIO).

In October 1946, Henry DeWolf Smyth, professor of physics wrote to Princeton president Harold Dodds, recommending Bohm as one of the ablest theoretical

[1] Among other sources, this section is based primarily on the study by Olwell (1999).

physicists turned out by Oppenheimer. Bohm joined Princeton as assistant professor in 1947 and started working on plasma physics, superconductivity, and quantum physics. While still in Princeton, on April 21, 1949, Bohm received a subpoena to testify before the House Committee on Un-American Activities (HUAC). This was to change Bohm's life for ever and also perhaps the development of quantum mechanics. Apparently, in the context of the cold war, Bohm's activities while in Berkeley now made him a security risk. Bohm took the Fifth Amendment when testifying before the HUAC, answering questions relating only to his background and education. He was brought to trial and finally acquitted of all charges on May 31, 1951. His colleagues and students at Princeton were generally supportive of his academic activities. Physics department chairman Allen G. Shenstone sent the following recommendation on November 15, 1950 to the Princeton president (Dodds): "The quality of his [Bohm's] scholarship is attested by the manuscript of a book on Quantum Mechanics which is based on his graduate course on the subject, and promises to be one of the outstanding books on the subject" (reproduced in Olwell, 1999, p. 747). President Dodds, a political scientist, however, thought that Bohm was a political liability, and consequently first suspended and later dismissed him from Princeton, and what is worse argued that this was due to lack of scholarly merits.

After leaving Princeton, it was difficult for Bohm to get another position, especially due to academic McCarthyism in the USA. In the summer of 1951 he joined the University of São Paulo, at the invitation of Brazilian physicists. His political problems continued, as the US Consul in São Paulo asked Bohm to return his passport, and that he could get it back only if he was returning to the US. Working conditions, both physical and political, were difficult in Brazil, which led to Bohm's isolation on the one hand and working primarily in theoretical physics. During these years, Bohm corresponded frequently with Einstein, who encouraged him both intellectually and morally. Einstein may have recalled his own experience in Nazi Germany and especially the role played by Philip Lenard (1862–1947), who had earlier worked on the photoelectric effect (see Chapter 8). Bohm, finally decided to give up his American citizenship for Brazilian, which enabled him to take up a job at the Technion, Haifa, Israel. Einstein approved Bohm's move to Israel and urged him to stay there. In 1957, Bohm moved to Bristol University in England and in 1961 received a chair in theoretical physics at Birkbeck College, University of London.

Origin and Development of Bohm's Interpretation of Quantum Mechanics

After leaving Princeton, Bohm devoted more time to understanding quantum mechanics and the realization that not all of the questions had been answered by the Copenhagen school, and later recalled in the following terms: "When I first taught quantum theory, I found that it was very unclear; and so I wrote a book which was mainly intended as an attempt to obtain a clear insight into Bohr's views on the subject" (Bohm, 1976, p. 4). This book was based on the graduate course that Bohm gave while he was at Princeton. In 1951, Bohm started working on his theory of "hidden variables" in quantum mechanics, which led to the publication in *Physical Review* (Bohm, 1952). This work brought Bohm into close contact with Einstein, then at the Institute for Advanced Study, and a leading critic of the Copenhagen school. Einstein, however, did not agree with the "hidden variables" interpretation. Bohm's major argument was that the probabilistic and the "indeterminacy" views of Bohr were the result of a failure to understand deeper laws and mechanisms, that he referred to as "hidden variables" (see next section for details).

Evaluation and reception of Bohm's work was inevitably enmeshed with his political views and the rhetoric of cold war debates. For example, some leading physicists (Jean Paul Vigier, Hans Freistadt, L. Janossy, and M. Schönberg) with Marxist leanings, rated Bohm's work very highly. The response of the American physics community in general was that of silence, neglect, or outright rejection. References to Bohm's (1952) major work (as reflected in *Science Citation Index*) remained low until the 1970s and started to increase thereafter, reaching to almost 50 in the early 1990s (cf. Olwell, 1999, Figure 3, p. 754).

At this stage it would be interesting to consider if Bohm's contemporaries ignored or critiqued his ideas based on the *milieu of the time* (cf. Nobel Laureate Leon Cooper, reproduced in Niaz et al., 2009). In other words, the sociopolitical circumstances of the Second World War and then the ensuing cold war did make development and later acceptance of Bohm's theory difficult. Recent resurgence of interest in Bohm's work provides evidence for this thesis. In contrast, Bohm himself was perhaps ambivalent by pointing out that he gained and lost at the same time (cf. Olwell, 1999, p. 755).

Comparison of Bohm's and Bohr's Copenhagen Interpretations of Quantum Mechanics

The standard or the orthodox view of quantum mechanics based on the Copenhagen interpretation (Bohr, Heisenberg, Pauli, and Born) is almost universally accepted by chemists, physicists, philosophers of science, and textbook authors. This interpretation requires complementarity (wave–particle duality), indeterminism, nonrealism, and the impossibility of an event-by-event causal representation in a continuous space-time background. A review of the literature shows that both physicists and philosophers of science are increasingly getting convinced with respect to alternative interpretations (Bohm, 1952) of quantum mechanics:

> Interest in Bohmian mechanics has rightly centered on its viability as an alternative interpretation of quantum mechanics. But it also offers calculation tools and physical insights that are unavailable in the standard approach. Although the heuristic aspect of Bohmian mechanics can facilitate understanding, and perhaps even discovery, it has evidently received scant attention. (Bowman, 2002, p. 313)

On the other hand, physicist-philosopher James Cushing has invoked underdetermination of scientific theories by experimental evidence (cf. Chapter 2, Duhem–Quine thesis, Quine, 1953), to argue for the importance of alternative theories:

> In a sense, Bohm's 1952 work can be seen as an exercise in logic – proving that Copenhagen dogma was not the only logical possibility compatible with the facts. In essence, Bohm accepted the formalism of quantum mechanics and showed that more microstructure is consistent with it than had previously been appreciated. By means of a reversible mathematical transformation, he was able to rewrite the Schrödinger equation in the form of Newton's second law of motion ("**F** = m **a**"). This result is exact and no approximations are made in obtaining it. In this theory, which predicts *all* of the results of standard quantum mechanics, there are event-by-event causality, a definite micro-ontology of actually existing particles that follow well defined trajectories in a space-time background – just the type of thing that was *in principle* forbidden by the Copenhagen interpretation! (Cushing, 1995, p. 139, original italics)

This may sound strange to many science students and teachers! The influence of the founding fathers of quantum mechanics (Bohr, Pauli, Heisenberg, and Born) facilitated not only the establishment of the Copenhagen hegemony, but also convinced the generation of physicists trained in the Copenhagen tradition to ignore Bohm's theory (cf. Cushing, 1996; Olwell, 1999).

A detailed account of Bohm's theory goes beyond the scope of this chapter. Nevertheless, a comparison of the "realism" in Bohmian mechanics and the "nonrealism" of Copenhagen quantum mechanics is presented. Physicist-philosopher Arthur Fine has facilitated a comparison of the classical, quantum mechanics (Copenhagen) and Bohmian views.

Classical Reality

> [I]n the classical picture what observation does is to reveal an observer-independent realm … [based on] the assumption that the observed-world is not very different from the unobserved world. How things are does not depend very much on whether things are observed. Observation is more passive than active. Reality is recorded not made. This picture of observation as disclosure is part of the "independence" that nature has from observation and it lies behind the idea that knowledge is objective. (Fine, 1996a, p. 244)

Copenhagen Reality

> According to quantum mechanics, however, the observer does make a difference, indeed a difference so big that one cannot really discount the effect of observations so as to frame an observer-independent picture of the world. This is a standard epistemological route to quantum nonrealism. (Fine, 1996a, p. 244)

Bohmian Reality

> He [Bohm] considered a wave function for the one dimensional motion of a "particle" moving back and forth between two perfectly rigid reflecting walls. Applied to a macro-particle, we know that, to a good approximation, it has a definite position and a (non-zero) velocity at (almost) each moment. Quantum mechanics does not yield that conclusion, and

> indeed no quantum state function will yield simultaneous approximate positions and momenta that are stable in the classical way. What quantum mechanics does provide, via Born's rule, are correct predictions for P or Q upon their respective measurement. Bohm's treatment of this example provides those same correct predictions, but also a picture of the situation before measurement.... According to the Bohmian prescription for velocity, the unmeasured particle is actually standing still! Measurement disturbs the situation, freeing the wave function, which guides the particle into motion. (Fine, 1996a, pp. 244–245)

At this stage the reader must be wondering: Why did the Copenhagen physicists (Bohr and others) hold fast to the notions of observer dependence and inevitable acausality, despite the existence of Bohm's alternative version? Beller (1999) has referred to a similar dilemma.

Interestingly, Bohm (1957) has himself responded to this question and attributes various reasons for this neglect:

(a) "[T]here appears to have existed, especially among those physicists such as Heisenberg and others, who first discovered the quantum theory, a rather widespread impression that the human brain is, broadly speaking, able to conceive of only two kinds of things, namely, fields [i.e., waves] and particles … [consequently] … only thing left to do will be to engage in purely technical manipulations of mathematical symbols according to suitable prescriptions" (Bohm, 1957, pp. 96–97).
(b) "[W]e should not postulate the existence of entities which cannot be observed by methods that are *already* available. This thesis stems from a general philosophical point of view containing various branches such as 'positivism', 'operationalism', 'empiricism', and others which began to attain a widespread popularity among physicists during the twentieth century" (Bohm, 1957, p. 97, original italics).
(c) In order to understand indeterminism, Bohm draws on the following analogy: "insurance companies operate on the basis of certain statistical laws, which predict to a high degree of approximation the mean number of people in a given class of age, height, weight, etc. that will die of a certain disease in a specified period of time. They can do this even though they cannot predict the precise time of death of an individual policy holder, and even though such individual deaths are distributed at random in a way having no lawful relationship to the kind of data that the insurance company is able to collect … in the field of medical research, the operation of statistical laws is never regarded as a reason against the search for more detailed individual laws (e.g., as to what makes a given individual die at a given time, etc.)" (Bohm, 1980, pp. 67–68).
(d) "Similarly, in the field of physics, when it was discovered that spores and smoke particles suffer a random movement obeying certain statistical laws (the Brownian motion) it was supposed that this was due to impacts from myriads of molecules, obeying deeper individual laws, for, as in the case of insurance statistics, the overall behaviour of an individual Brownian particle would be determined by a very large number of essentially independent *factors* [hidden variables]" (Bohm, 1980, p. 68, emphasis added).
(e) Returning to quantum mechanical measurements and based on a historical reconstruction, Bohm concludes that "These *factors* would be represented mathematically by a *further set of variables*, describing the states of new kinds of entities in a deeper, sub-quantum-mechanical level and obeying qualitatively

new types of individual laws. Such entities and their laws would then constitute a new side of nature, a side that is, for the present 'hidden'. But then the atoms, first postulated to explain Brownian motion and large-scale regularities, were also originally 'hidden' in a similar way, and were revealed in detail only later by new kinds of experiments (e.g., Geiger counters, cloud chambers, etc.) that were sensitive to the properties of individual atoms" (Bohm, 1980, pp. 68–69, emphasis added).

Bohm's historical reconstruction is indeed illuminating and indicates quite clearly as to how in the history of science the quest for further and deeper understanding through as yet unknown "hidden" variables has been a source of controversy and conflict. It is worthwhile to recall the role played by atomic theory in understanding Brownian motion and the opposition by some of the leading scientists (Mach and Ostwald) in the late nineteenth and early twentieth centuries. To conclude, it appears that according to Cushing's (1991) scheme, Copenhagen interpretation works preferably at the level of explanation (formalism of quantum mechanics), whereas Bohmian mechanics attempts to facilitate understanding.

Educational Implications of the Historical Reconstruction

Niaz and Fernández (2008) have reported that of the 55 general chemistry textbooks (all published in the USA) analyzed, none referred to the contribution of Einstein towards the origin of the quantum hypothesis. The objective of this criterion was to evaluate textbooks' presentation of the origin of the quantum hypothesis, based primarily on the distinction between Planck's contribution as a mathematical or calculation device to understand black-body radiation and Einstein's contribution as to the physical significance of the postulate. Interestingly, Styer (2000, p. 3) considers the concept of threshold energy (Einstein's photoelectric effect) as crucial for understanding the difference between quantum and classical mechanics, as the latter would precisely predict that any wavelength of light could eject an electron. To conclude, the degree to which Planck did not explain the idea of threshold energy and the almost instantaneous appearance of photoelectric current are reasonable grounds for assuming his theory to be more classical. By emphasizing these differences between Planck's and Einstein's contributions, textbooks can facilitate greater conceptual understanding.

Textbooks were also analyzed on the following criterion: based on the underdetermination of scientific theories by experimental evidence it is plausible to suggest that besides the standard Copenhagen interpretation of quantum mechanics there are alternatives equally compatible with experimental facts, such as Bohm's interpretation based on "hidden variables." The latter interpretation is particularly helpful in facilitating greater physical insight in atomic structure. Five textbooks made a simple mention and following is an example: "The uncertainty principle is not easy for most people to accept. Einstein spent a good deal of time from the middle of the 1920s until his death in 1955 attempting, unsuccessfully, to disprove it"

(Petrucci & Harwood 1997, p. 285, on page 309 the student is provided an exercise in which he is asked to comment on Einstein's famous statement about God playing dice). None of the textbooks referred to Bohm's interpretation of quantum mechanics. One textbook did provide an alternative based on "The Dirac equation: A relativistic model of the atom" (Umland & Bellama, 1999). This aspect is all the more important as atomic theory itself when first postulated was also based on "hidden" variables and led to considerable controversy.

Recognition of alternative interpretations of quantum mechanics can help students to understand that scientific theories are in general underdetermined by experimental evidence. Bohm (1952, 1957, 1980) provides interesting examples (see section on A Brief Introduction to the Life and Career of David Bohm) from the history of science (cf. atomic theory and Brownian motion) to justify his "hidden variables" interpretation. A recent attempt to introduce Bohmian mechanics at the undergraduate level emphasized precisely the physical understanding: "By providing a valuable link with familiar classical physics, the Bohmian approach arguably can provide a better understanding of *why* quantum systems behave as they do. This understanding, in turn, can help develop physical intuition – an elusive commodity in the quantum world" (Bowman, 2002, p. 317, original italics).

Chapter 12
Wave–Particle Duality: De Broglie, Einstein, and Schrödinger

Introduction

Postulation and understanding of wave–particle duality was a controversial topic from the very beginning and is closely enmeshed with the origin and development of the photoelectric effect and quantum theory. De Broglie (1924) in a seminal paper explored the reconciliation of lightquanta (used by Einstein to explain the photoelectric effect) with "the strong experimental evidence on which was based the wave theory" (p. 446). J.J. Thomson (1925), in his *Structure of Light*, compared the interplay between wave and particle theories of radiation to a struggle between a tiger and a shark in which each is supreme in his own element, but helpless in that of the other. Millikan (1916) provided experimental evidence for Einstein's photoelectric equation and still rejected the hypothesis of light quanta. Actually, Millikan went far beyond this by considering Einstein's hypothesis reckless, as it could not explain thoroughly established facts of interference. Millikan, however, conceded that as the photoelectric effect could not be explained by the classical theory, it may at the most need some modifications and not its rejection. Millikan's example in this case is a good indicator of how prior epistemological beliefs of the scientists play an important role in the acceptance of new ideas. In the present case, even as late as 1924 Millikan believed that the classical wave theory only needed to be reinterpreted. Lakatos (1970) has explained that in the history of science, scientists frequently resist changes in the "hard core" of their theories.

A historical reconstruction of wave–particle duality, photoelectric effect, and quantum theory has been recognized in cogent terms by Noyes (1984), a physicist:

> The foundations of quantum physics were laid between 1896 and 1925.... Most practicing physicists have learned what little they know of the history of this period by reading textbooks written after the quantum revolution. Often texts or teachers treat the Planck radiation law, the Einstein photoelectric equation, the Bohr atom and the Compton effect in one sequence assuming that this provides an adequate background for understanding $E = h\nu$ and $p = h\nu/c$. This can leave a student with less than total respect for the physicists who took so long to see the "obvious" necessity for this form of quantization. (p. 95)

The objective of this chapter is to present a historical reconstruction of the wave–particle duality and the role played by de Broglie, Einstein, and Schrödinger, which is presented in the following sections:

M. Niaz, *Critical Appraisal of Physical Science as a Human Enterprise: Dynamics of Scientific Progress*, Science and Technology Education Library 36,

1. Origin of the wave–particle duality
2. De Broglie and wave–particle duality
3. De Broglie's reputation as an obstacle in the acceptance of his theory
4. Einstein's support of de Broglie's ideas
5. Why was it Schrödinger who developed de Broglie's ideas?

Origins of the Wave–Particle Duality

Hendry (1980) has suggested that the origins of wave–particle duality and its relationship to quantum theory can be studied through the following stages: *Stage 1 (intimations of duality), 1900–1908*: by 1908, Planck's radiation law was well established, but its significance was still not established. Based on Planck's authority, physicists still recognized Philip Lenard's trigger hypothesis to explain the photoelectric effect (cf. Wheaton, 1978) and generally rejected Einstein's (1905) light quantum hypothesis. However, it was becoming clear that Planck's law contained the essence of the wave–particle duality; *Stage 2 (recognition of a paradox), 1909–1913*: working on the assumption that the classical theory of radiation needed to be radically revised, Einstein (1909a, b) published two important papers in 1909, in which he argued explicitly for a fusion of the wave and emission theories of light. This support for the wave–particle duality, however, was not accepted by Planck, Wien, Sommerfeld, and others. Einstein's advocacy of duality was even seen as a paradoxical retreat from the idea of light quanta of 1905, back toward the classical view. Even in 1912, Planck's main argument was that the quantization should apply to the action in an atomic process rather than to the energy of emitted or absorbed radiation. Similarly, Bohr (1913a) did not accept this discreteness as strictly applying to the radiation. There was, however, one new positive development when J.H. Jeans at the 1913 meeting of the *British Association for the Advancement of Science* announced his support for the wave–particle duality; *Stage 3 (impasse), 1914–1920*: Millikan (1916) provided experimental evidence for Einstein's photoelectric law, but still rejected the light quantum hypothesis (see Chapter 8). Einstein's (1916) efforts to present a new derivation of Planck's law based on a new notion of absorption and emission probabilities still remained inconclusive. Wave–particle duality was once again a subject of discussion at the Third Solvay Conference in 1921.

De Broglie and Wave–Particle Duality

De Broglie's (1922) was the first attempt to study black-body radiation in the context of lightquantum (Medicus, 1974, p. 38). Later, this interest in the properties of quanta motivated de Broglie's search for a theory that would unify the wave and particle aspects. In 1923, he published three short articles in *Comptes Rendus* in which he generalized wave–particle duality to include material corpuscles (de Broglie, 1923a). This was followed by a short note to *Nature* (de Broglie, 1923b) and then the complete article in *Philosophical Magazine* (de Broglie, 1924).

De Broglie's First Attempt to Postulate Wave–Particle Duality

De Broglie (1923a) applied wave–particle duality hypothesis to existing problems in physics and among others he referred to the following important issues (cf. Medicus, 1974, p. 40):

(a) Application of his hypothesis to electron orbits in an atom, requiring that the wave be in phase with itself, and that the circumference be an integral multiple of the wavelength. In the summary of his article he concluded: "We believe that this is the first physically plausible explanation for the Bohr–Sommerfeld stability rules" (reproduced in Medicus, 1974, p. 40).
(b) The famous formula $\lambda = h/mv$ is found in this explicit form for the first time in the chapter on statistical mechanics. For de Broglie it is not the wavelength of the particle but the frequency that is more important.
(c) There is no essential difference between photons and particles. De Broglie assumed that light quanta have an extremely small mass of the order of 10^{-50} g.
(d) De Broglie even suggested a possible experimental confirmation of his hypothesis in the following terms: "A beam of electrons passing through a very small opening could present diffraction phenomena. This is perhaps the direction in which one may search for an experimental confirmation of our ideas" (reproduced in Medicus, 1974, p. 40). The note to *Nature* (de Broglie, 1923b) also referred to this possibility.

Experimental Evidence to Support de Broglie's Theory

It is interesting to note that de Broglie's theory of wave–particle duality not only preceded its experimental confirmation, but he also suggested how experiments could be performed. In 1968, in an interview with F. Kubli (1970), de Broglie said that his brother Maurice de Broglie had suggested that the theory should also include an experimental part. De Broglie declined, saying that he was not an experimentalist. During his doctoral examination in November 1924 at the Sorbonne, J. Perrin, chairman of the examination committee, asked how one could experimentally observe matter waves (cf. Medicus, 1974, p. 40). De Broglie once again suggested diffraction experiments on crystals with electrons. In a letter written to H.A. Medicus on September 17, 1973, de Broglie said that at the time of his thesis he had suggested A. Dauvillier (an astrophycicist) to undertake experiments to obtain diffraction or interference phenomena with electrons. Due to experimental difficulties, these efforts were unsuccessful. Furthermore, in a letter written to H.A. Medicus on November 16,

1973, Dauvillier reported that "it is indicative that neither M[aurice] de Broglie, nor P. Langevin, nor J. Perrin saw to it that [such experiments] were carried out in their laboratories! Nobody believed in it" (reproduced in Medicus, 1974, p. 40). In 1952, C. Maugin (a crystallographer and a member of de Broglie's doctoral examination committee) recalled: "Today I have difficulty understanding my state of mind [in 1924] when I accepted the explanation of the facts [the Bohr-Sommerfeld quantization rules] without believing in the physical reality of the entities that provided this explanation" (reproduced in Medicus, 1974, p. 41). Interestingly, Medicus (1974) after this comment noted: "Today, … we have no such inhibitions about accepting the quark model, for example, even though no one has ever found a quark!" (p. 41). This reconstruction shows how theories develop and the intricate relationship between physical reality and experimental determination. Martin Perl, a Nobel Laureate in physics, has been working for the last several years on the isolation of quarks (see Chapter 13) and has formulated a philosophy of speculative experiments for experimental confirmation: "Choices in the design of speculative experiments usually cannot be made simply on the basis of pure reason. The experimenter usually has to base her or his decision partly on what feels right, partly on what technology they like, and partly on what aspects of the speculations they like" (Perl & Lee 1997, p. 699). From C. Maugin's belief of physical reality to M. Perl's philosophy of speculative experiments, we can understand how scientific research methodology has evolved over the last 7 decades (Rodríguez & Niaz, 2004b).

Walter Elsasser at the University of Göttingen was influenced by Einstein and de Broglie's ideas about wave–particle duality. On reading de Broglie's theory, Elsasser (1925) reinterpreted the previously published results of experiments performed by Davisson and Kunsman (1923), as evidence for the wave nature of the electrons. These experiments dealt with scattering of slow electrons from metal surfaces, so that when deflected from different shells of the atoms, electrons would be deflected in different ways and thus served as probes to atomic structure. On reading Elsasser's article, Einstein told him: "Young man, you are sitting on a gold mine!" (reproduced in Medicus, 1974, p. 43). Davisson, however, did not agree with Elsasser's interpretation of his experimental results as evidence for wave–particle duality (Davisson & Germer, 1927, p. 707). Elsasser on his part wanted to continue the experiments, but was denied support by James Franck, head of the Institute at Göttingen. Elsasser, then switched to theoretical physics under the direction of M. Born.

Davisson continued with his experiments and an accident in his laboratory provided unexpected lead to the understanding of de Broglie's theory. In their seminal article, Davisson and Germer (1927) started by recounting the accident: "The investigation reported in this paper was begun as the result of an accident which occurred in this laboratory in April 1925.… During the course of this work a liquid-air bottle exploded at a time when the target was at a high temperature; the experimental tube was broken, and the target heavily oxidized by the inrushing air. The oxide was eventually reduced and layer of the target removed by vaporization, but only after prolonged heating at various high temperatures in hydrogen and in vacuum. When the experiments were continued it was found that the distribution-in-angle of the scattered electrons had been completely changed" (p. 706). This change was attributed to a recrystallization of the target into a few relatively large single crystals during the heating process. Interpretation of the data, however, was not straightforward. A scientific meeting at Oxford in August 1926 helped Davisson to consult M. Born and other physicists that brought him onto the right track (Medicus, 1974, pp. 43–44). Davisson and Germer made a preliminary announcement of their results in *Nature*, April 16, 1927 and then published the complete report in *Physical Review* in December 1927 and concluded:

> Because of these similarities between the scattering of electrons by the crystal and the scattering of waves by three- and two- dimensional gratings a description of the occurrence and behavior of the electron diffraction beams in terms of the scattering of an equivalent wave radiation by the atoms of the crystal, and its subsequent interference, is not only possible, but most simple and natural. This involves the association of a wave-length with the incident electron beam, and this wave-length turns out to be in acceptable agreement with the value *h/mv* of the undulatory mechanics. (Davisson & Germer, 1927a, p. 707)

George P. Thomson (son of J.J. Thomson), while at the University of Aberdeen read de Broglie (1924) and like Davisson also attended the scientific meeting at Oxford in 1926. In contrast, Thomson worked with electron diffraction of solids (celluloid, gold, aluminum, and platinum), which permitted him to use electrons of considerably higher energies, and observed diffraction rings. A preliminary announcement of his results was made in *Nature* in June 1927 (Davisson did in April). The complete report was published in the *Proceedings of the Royal Society* in February 1928 (Thomson & Reid, 1928). Both Davisson and Thomson shared the 1937 Nobel Prize in physics.

De Broglie's Reputation as an Obstacle in the Acceptance of His Theory

Contrary to popular belief, de Broglie was not a young, unknown student when he wrote his major paper on wave–particle duality in 1923. He was 32 years old when he received his doctorate and before that worked for 6 years in the military, stationed at the radiotelegraphic station on the Eiffel tower. Before his doctorate he had published about 2 dozen scientific papers on electron, atomic, and x-ray physics. It was precisely this earlier work that led him into various controversies (especially Bohr's Copenhagen school) and also provided resistance to the acceptance of his novel wave–particle duality. Some of these controversies are summarized below (complete details are available in Raman & Forman, 1969): (a) in 1921, although de Broglie accepted Bohr's correspondence principle, he gave it a different interpretation, which was critiqued by H.A. Kramer (a colleague of Bohr); (b) question of whether or not every degree of freedom of every electron in an atom is entitled to a quantum number. Bohr and most theoretical atomic physicists disagreed, whereas de Broglie responded in the affirmative; (c) question of the number of x-ray and optical energy levels associated with each value of the principal quantum number; (d) number of electrons with given quantum numbers in the atom of each element; (e) Copenhagen school claimed to have isolated element 72 in 1923 and proposed the name of Hafnium. Based on the work of A. Dauvillier in 1922, de Broglie proposed the name "celtium"; (f) de Broglie also got into a controversy in 1924 with A. Sommerfeld's (Munich school) procedure in applying the quantum conditions to electronic orbits, which was aggravated besides by French–German animosities. According to Raman and Forman (1969), just when de Broglie presented his theory of wave–particle duality, his scientific reputation was questioned: "Thus in Copenhagen – and in Göttingen, where atomic physics was pursued in the Copenhagen spirit – de Broglie would certainly have had the reputation of a renegade, if not exactly a crank, who stuck obstinately to his own ill-conceived theories" (p. 295).

Einstein's Support of de Broglie's Ideas

Despite de Broglie's reputation, Einstein was a strong supporter of wave–particle duality from the very beginning. Langevin (de Broglie's thesis advisor) sent a copy of the thesis to Einstein before the examination (November 1924). Einstein wrote back to Langevin saying de Broglie had "lifted a corner of the great veil" (reproduced in Medicus, 1974, p. 41). In December 1924, Einstein was working on a paper in which he calculated the fluctuations of an ideal gas (Bose–Einstein statistics) and he incorporated de Broglie's ideas (published in the *Proceedings of the Prussian Academy* in February 1925, Einstein, 1925), with respect to association of a wave field with a material particle. Einstein showed how de Broglie's $m_0c^2 = h\nu$ necessarily leads to a wave, exposes the relation between phase and group velocity, and the fact that the group velocity of the wave is the velocity of the particle. Furthermore, he drew attention to an important geometrical interpretation of the Bohr–Sommerfeld quantization rule. Einstein's (1925) support and use of de Broglie's ideas influenced at least two other physicists: Walter Elsasser in Göttingen and Erwin Schrödinger in Zurich (p. 41, Medicus, 1974). In his paper on gas theory, Schrödinger (1926a) noted: "This means nothing else but to take seriously the de Broglie–Einstein undulation theory of moving corpuscles" (p. 95). Later in his major paper on the relationship between wave and matrix mechanics, Schrödinger (1926b) explicitly acknowledged that his theory was inspired by de Broglie's ideas. These events set the stage for Schrödinger to develop de Broglie's ideas (cf. Darrigol, 1993).

Why Was It Schrödinger Who Developed de Broglie's Ideas?

De Broglie's thesis on wave–particle duality was given to Schrödinger by V. Henri, but returned shortly with the remark: "That's rubbish" (reproduced in Raman & Forman, 1969, p. 311, also Jammer, 1966 p. 258). Langevin on knowing this instructed Henri to insist that Schrödinger have another look at the thesis. According to Raman and Forman (1969) "even Schrödinger, the man who was in some way fated to develop de Broglie's ideas into a wave mechanics, did not recognize these ideas as promising when first exposed to them" (p. 311).

Given de Broglie's reputation, theoretical spectroscopists (primarily Copenhagen and Munich schools) were not likely to invest much time and effort in trying to make sense of phase waves associated with material particles. It is plausible to suggest that Schrödinger was particularly suited to develop de Broglie's ideas for the following reasons: (a) Schrödinger, a marginal man, a loner, was not a member of Copenhagen/Munich schools of spectroscopists, and still made important contributions to theoretical spectroscopy; (b) Einstein's early endorsement of de Broglie's theory was crucial for Schrödinger, which he acknowledged explicitly in a letter to Einstein dated April 23, 1926: "Your and Planck's assent are more valuable to me than that of half the world. Moreover, the whole business [wave mechanics] would certainly not yet, and perhaps never, have been developed (I mean not by me) if the importance of de Broglie's ideas hadn't been put right under my nose by your second paper on gas degeneracy" (reproduced in Raman & Forman, 1969, p. 311; second paper on gas degeneracy refers to Einstein, 1925); (c) Schrödinger (1922) had previously published ideas, which were conceptually similar (although not identical) to de Broglie's (cf. Raman & Forman, 1969, p. 310); (d) Schrödinger's involvement in the fundamental problems of quantum statistical mechanics (Klein, 1964); (e) conceptually, Schrödinger aligned himself with Einstein in his controversy with the Copenhagen school of quantum mechanics on important issues such as causality, probability, and determinism.

At this stage it is important to recall that de Broglie's theory and other aspects of quantum mechanics were the subject of extensive discussions at the 1927 Solvay Conference in Brussels. Present at the conference were some of the leading physicists, like Einstein, Bohr, de Broglie, Schrödinger, Heisenberg, Pauli, Born, Lorentz, among others. The famous Einstein–Bohr debate and thought experiments designed by Einstein to refute the Copenhagen interpretation started at this conference. A recent commentator looking back at the conference has analyzed the possible reasons for the "landslide" acceptance of Copenhagen interpretation after the conference (Bonk, 1994). Among other reasons, Bonk has attributed this to a working "positivist" theory in the Copenhagen framework and defunct "realist" alternatives including de Broglie's theory. Despite the rejection of de Broglie's theory, Bonk (1994) has asked a very pertinent question: "Now, would it have been 'irrational' for a scientist to believe, to investigate and develop de Broglie's alternative to the Copenhagen interpretation?" (p. 393). Interestingly, Bonk (1994) has himself responded that "judgments in de Broglie's case were based prematurely on a not fully understood and ill-defended theory" (p. 393).

The importance of historical reconstructions of various topics for general chemistry and physics textbooks has been recognized in the science education literature (Niaz, 1998, 2000a, b; Niaz et al., 2005; Niaz & Fernández, 2008; Rodríguez & Niaz, 2004a, b). It is plausible to suggest that the historical reconstruction of the wave–particle duality presented in this chapter can have implications for textbooks and teaching this topic in introductory physical science courses.

Chapter 13
Searching for Quarks: Perl's Philosophy of Speculative Experiments

Introduction

Although most science courses and textbooks emphasize that the fundamental, indivisible unit of electric charge is the electron, physicists have been searching for fractional charges since the 1960s. Since the initial measurements of the electron charge a century ago, experimenters have faced the persistent question as to whether elementary particles exist that have charges that are fractional multiples of the electron charge. Millikan himself in his Nobel Prize acceptance speech in 1923 envisioned: "If the electron is ever subdivided it will probably be because man, with new agencies as unlike X-rays and radioactivity as these are unlike chemical forces, opens up still another field where electrons may be split up without losing any of the unitary properties which they have now been found to possess in the relationships in which we have thus far studied them" (Millikan, 1965, p. 61).

The objective of this chapter is to present Martin Perl's (Nobel Laureate in physics, 1995) strategy of speculative experiments in his search for quarks and its implications for understanding the Millikan–Ehrenhaft controversy (oil drop experiment, see Chapter 7) and cutting-edge experiments. This reconstruction is presented in the following sections:

1. Search for elementary particles with fractional electrical charge
2. Understanding scientific research methodology: contrasting Millikan and Perl
3. Conclusion: the role and importance of cutting-edge speculative experiments

Search for Elementary Particles with Fractional Electrical Charge

One of the few and first studies that reported the finding of quarks (fractional charges) was by LaRue et al. (1981). Soon after this, Fairbank and Franklin (1982) reanalyzed Millikan's (1913) original data to reexamine the evidence for charge quantization

M. Niaz, *Critical Appraisal of Physical Science as a Human Enterprise: Dynamics of Scientific Progress*, Science and Technology Education Library 36,

(unitary electron) versus fractional charges (subelectrons): "Recent reports [LaRue et al., 1981] of the observation of fractional residual charges on super conducting niobium spheres have aroused interest in re-examining Millikan's original oil drop data" (p. 394). This shows the relationship between the Millikan–Ehrenhaft controversy and the finding of fractional charges (quarks) in recent research (cf. Chapter 7 for details about the controversy). Although, Millikan and Ehrenhaft could not have referred to quarks in their controversy (1910–1925), they did deal with fractional charges. The controversy actually was based on Ehrenhaft's finding of drops with fractional charges (subelectrons) and Millikan's evidence for the unitary electron (cf. Dirac, 1977; Holton, 1978; Pickering, 1984). A recent appraisal clarifies the issue in the following terms: "In 1977, after several years of work, the Stanford group reported that it had found fractional charges in Millikan-like experiments, namely, three cases of tiny niobium balls carrying one-third of the electron's charge. The claim was controversial both experimentally and theoretically, for at that time, theorists had become convinced that free quarks do not exist but are forever confined within hadrons of which they are parts" (Kragh, 1999, p. 324). In other words, work on quarks has been associated with fractional charges, and students could benefit from a discussion of the work on quarks in the context of the oil drop experiment. This could also illustrate how a controversy in the past can manifest in a different experimental situation.

Early searches for quarks were based on the following assumptions (Jones, 1977): (a) they had fractional electric charge (±1/3, ±2/3); and (b) they could be produced if sufficient energy was provided either to dissociate an energetic hadron into its quark constituents or to produce them in pairs. According to Pickering (1984): "From the experimental point of view, quarks had one key attribute: fractional electric charge. The belief had come down from the days of Robert Millikan that all matter carried charges in integral multiples of the charge on the electron. Now this belief had been challenged, and experimenters were not slow to respond" (p. 88).

The concept of fractional charge elementary particles is tightly connected with our present understanding of the nature of quarks. According to the consensus Gell–Mann–Zweig assignment, quarks have fractional charge f equal to 2/3 or −1/3 and the antiquarks have the opposite sign charges.

Perl and Lee (1997) have presented a critical review of attempts to search and isolate fractional charges:

1. *Accelerator experiments*: In this technique (also referred to as high-energy physics) the available energy puts an upper limit on the mass of the fractional charge. Furthermore, the fractional charge "may interact and stop in the walls of the accelerator, target, or detecting apparatus, and hence it may be produced but not found" (Perl & Lee, 1997, p. 699).
2. *Cosmic rays*: According to this research, the fractional charge may be contained in the cosmic ray flux coming from outside the earth's atmosphere or might be produced in a cosmic ray collision in the earth's atmosphere. Once again the available energy puts an upper limit on the detection of fractional charges.
3. *Bulk matter*: According to this research, fractional charges may have been produced in the early universe and come to earth, perhaps during the formation

of the earth or later by meteorite landing. The sensitivity of this research depends in part upon the amount of bulk matter examined. According to Perl and Lee (1997): "If we ignore the claims of LaRue et al. [1981] to have found *f*'s [fractional charges], then no *f*'s have been found in material containing about 10^{21} nucleons" (p. 702).

4. *Automated Millikan liquid drop method*: This is the method favored by Perl and Lee (1997) and various other investigators (Mar et al., 1996). Interestingly, besides incorporating modern technology (piezoelectric drop generators, digital cameras, image processing electronics, high speed computers, etc.), their method remains quite similar to that of Millikan's, and they explicitly cite Millikan (1910, 1911, 1917). Perl and Lee (1997) have summarized this in the following terms: "This was Millikan's method and this is our method eighty years later. Millikan studied a few hundred drops, we have studied almost 10^7 drops and will soon study 10^8–10^{10}. Our improvement over Millikan is made possible by the use of modern technology" (p. 700). Although these investigators have so far not been able to isolate fractional charges, they recognize that the potential for finding such charges has increased considerably. These findings have two implications: (a) modern techniques have vindicated Millikan's controversial methodology (cf. Niaz, 2000b, 2005b for the controversy with Ehrenhaft, also Chapter 7); and (b) Millikan and Ehrenhaft may not have observed fractional charges (quarks) due to the lack of appropriate modern techniques. Dirac (1977), after studying the experimental conditions under which Ehrenhaft and Millikan worked, concluded: "This does not constitute evidence for quarks. It merely shows there was some experimental error, perhaps the same for both of them [Ehrenhaft and Millikan], affecting their smaller particles" (p. 293).

More recently, Perl et al. (2004, p. 14) have reviewed research related to isolating quarks and suggested three alternative possibilities:

1. Although a very large number of searches for isolatable fractional charge particles have been carried out using a variety of techniques, there is no confirmed evidence for the existence of such particles. Perhaps there are *no* isolatable fractional charge particles. Indeed almost all physicists seem to have come to this conclusion as demonstrated by the very small interest by experimenters these days in searching for isolatable fractional charge particles.
2. Isolatable fractional charge particles do exist, but the masses of fractional charge particles are so large or that their production mechanisms are so small that they were not produced in the early universe. Furthermore, they cannot be produced now by ongoing natural processes in the universe or by existing accelerators or colliders.
3. To recognize that the searches so far carried out, while extensive, were bounded by the technique and the technology. Consequently, with new technology and inventions, the range of search parameters can be extended. Perl et al. (2004) consider this to be the most plausible state of research related to isolating quarks and suggest the invention of ways to substantially increase the range of known search methods.

Understanding Scientific Research Methodology: Contrasting Millikan and Perl

Given the difficulties of isolating quarks, Perl and Lee (1997) have suggested to work outside the present experimental boundaries established by quantum chromodynamics. Such efforts involve both experimental and speculative aspects, which has led them to propose a philosophy of speculative experiments. In order to justify their methodology, the authors have made eight philosophical remarks. It appears that these remarks can be sustained by referring to the research methodology of Millikan in the context of the oil drop experiment. Furthermore, it helps to elucidate the controversy with Ehrenhaft and show that in such experimental work the traditional scientific method is not very helpful.

In this section we present the eight philosophical remarks of Perl and Lee (1997) along with different aspects of Millikan's methodology, in order to show how it constitutes an illustration of scientific research methodology. It is expected that this illustration can enrich our present-day understanding of scientific research methodology.

Philosophical Remark 1 (Perl & Lee, 1997)

In sensible, speculative experimental work, the research must extend beyond established experimental knowledge (p. 698).

Millikan's work extended the work previously done by Townsend (1897), Thomson (1898), and Wilson (1903) on charged clouds of water droplets. Millikan explicitly critiqued the Townsend's and Thomson's studies for not taking into account the evaporation of water droplets (Millikan & Begeman, 1908). For details, see Niaz (2000a, pp. 483–484).

Philosophical Remark 2 (Perl & Lee, 1997)

Choices in the design of speculative experiments usually cannot be made simply on the basis of pure reason. The experimenter usually has to base his/her decision partly on what feels right, partly on what technology they like, and partly on what aspects of the speculations they like (p. 699).

In his controversy with Ehrenhaft, Millikan's arguments were not precisely based on experimental evidence, but rather what he felt was the correct value of the elementary electrical charge: "That these same ions have one sort of charge when captured by a big drop and another sort when captured by a little drop is obviously *absurd*.... such an assumption is not only too *grotesque* for serious consideration but is directly contradicted by my experiments" (Millikan, 1916, p. 617, emphasis added). Although, Millikan does allude to his experiments, it is important to note that Millikan discarded many (more than half in some cases) drops that did not have a charge equivalent to a multiple of the elementary electrical charge. A novice would ask: Can experimental data be referred to as "absurd" or "grotesque?" For details, see Niaz (2000a, pp. 490–491). Polanyi (1964) provides helpful suggestions with respect to this dilemma: "Our vision of reality, to which our sense of scientific beauty responds, must suggest to us the kind of questions that it should be reasonable and interesting to explore. It should recommend the kind of conceptions and empirical relations that are intrinsically plausible and which should therefore be upheld, even when some evidence seems to contradict them, and tell us also, on the other hand, what empirical connections to reject as specious, even though there is evidence for them – evidence that we may as yet be unable to account for on any other assumptions" (p. 135).

In a recent study we asked Leon Cooper (Nobel Laureate, physics, 1972) to comment on Philosophical Remark 2 of Perl and Lee (1997). Cooper endorsed this remark in the following terms:

> Of course Perl is right. Pure reason is great. Experimentalists base their decision of what experiments to do on what feels right, what technology they're capable of using and their intuition as to what can be done and what might really be an important result. Experimentalists sometimes say that the first thing they try to do in an experiment is to make it work. It is intuition guided by facts, conjectures, and thoughts about what really would be important. (Reproduced in Niaz et al. 2009)

This makes interesting reading, as Cooper in a sense goes beyond Perl and Lee, by emphasizing not only speculations, but also intuition guided by facts and conjectures.

Philosophical Remark 3 (Perl & Lee, 1997)

A beautiful experiment is always pleasant, but beauty is very desirable in speculative research where beauty may be the only reward (p. 699).

Millikan's procedure seems to have consisted of making a rough calculation for the value of *e*, as soon as the data for the times of descent/ascent of the oil drops started coming in. In his laboratory notebooks (October 28, 1911–April 16, 1912) Millikan wrote comments along with the data, such as: "*Error high* will not use … can work this up and probably is ok but point is [?] not important. Will work if have time" (reproduced in Holton, 1978, p. 209). On another occasion Millikan wrote: "Beauty. *Publish* this surely, *beautiful!*" (p. 209). A novice may ask: How does a scientist know as to what data can or should be published? According to Polanyi (1964): "The affirmation of a great scientific theory is in part an expression of delight. The theory has an inarticulate component acclaiming its beauty, and this is essential to the belief that the theory is true" (p. 133).

Philosophical Remark 4 (Perl & Lee, 1997)

Choose the most modern technology for a speculative experiment, particularly the technology that is continually getting better and cheaper. Then you will not be frustrated by your speculations exceeding your technological reach (p. 700).

Besides other innovations (e.g., oil instead of water) Millikan used an exceptionally large battery of 10,000 V (compared to 4,000 V in previous experiments), which dispersed the cloud of water droplets instantaneously and left a small number of individual drops. This is precisely why it is referred to as the oil drop experiment. Millikan (1950) recalled later in his autobiography: "[the dispersal] seemed at first to spoil my experiment. But when I repeated the test, I saw at once that I had something before me of much more importance than the top surface [of cloud] … For repeated tests showed that whenever a cloud was thus dispersed by my powerful field, *a few individual droplets would remain in view*" (p. 73, original italics). According to Holton (1978, p. 183), this brought the decade-long technique of measuring electrical charges by the formation of clouds of water droplets to an abrupt end.

Philosophical Remark 5 (Perl & Lee, 1997)

In a speculative research it is wise to minimize experimental restrictions (p. 700).

It is important to note that Ehrenhaft too obtained data that he interpreted as integral multiple of the elementary electrical charge, *e* (similar to Millikan). Nevertheless, he was not willing to discard drops (and hence minimize experimental restrictions) that did not lead to an integral multiple of *e*. Indeed, Millikan would perhaps have liked to warn Ehrenhaft that all the readings cannot be used as their experiments were constantly faced with difficulties such as evaporation, sphericity, radius, and change in density of oil drops and variation in experimental conditions (battery voltages, stopwatch errors, temperature, pressure, convection, and so on).

Philosophical Remark 6 (Perl & Lee, 1997)

In speculative experiments there is generally not a clear best way, experimenters are free to follow their hearts as well as their heads (p. 702).

The Millikan–Ehrenhaft controversy is a good illustration of how there was "not a clear best way" of determining the elementary electrical charge. Furthermore, if the "best way" (considered to be the traditional scientific method by most textbooks and even some practicing scientists) had been followed, the scientific community would have accepted Ehrenhaft's experimental data. Interestingly, a general chemistry textbook described Millikan's work in the following matter of fact way: "The very large value of e/m could result either from a large value of the electronic charge, or from a small value of the mass of the electron. An independent measurement of either *e* or *m* was therefore necessary, and in 1910 Robert A. Millikan determined the charge on the electron in a classic and famous experiment" (Segal, 1989, p. 412). Such presentations do not facilitate student understanding of scientific progress. Stinner (1989) refers to this as the context of questions, viz., "how presuppositions produce questions whose answers then shape the particular science, and determine what counts as scientific fact" (p. 594).

Philosophical Remark 7 (Perl & Lee, 1997)

In speculative experimental science, one researcher's uncertainty is another researcher's hope and opportunity (p. 702).

Ehrenhaft by including data from all the drops made the interpretation of his results difficult and hence uncertain. On the contrary, Millikan by discarding oil drops that did not support the existence of the elementary electrical charge (*e*) ceased the opportunity to gain support for an alternative hypothesis. For details see Holton (1978) and Niaz (2000a).

Philosophical Remark 8 (Perl & Lee, 1997)

If a speculative experiment succeeds there will be many doubters, therefore, it is important that the experimental apparatus be easy to duplicate so that the unexpected discovery can be verified by others (p. 703).

In the Millikan–Ehrenhaft controversy, support for Millikan's position was not spontaneous and many leading physicists withheld judgment. H.A. Lorentz in the 1916 edition of *The Theory of Electrons* stated: "Millikan has found values for *e* which can be considered as multiples of a definite 'elementary' charge. Ehrenhaft, however, has been led to the conclusion that in some cases the charges are not multiples of the elementary one and may even be smaller than it. The question cannot be said to be wholly elucidated" (Lorentz, 1952, p. 251). Although in 1923 Millikan was awarded the physics Nobel Prize, as late as 1922, R. Bär in a review of the controversy recognized: "The experiments [Ehrenhaft's] left, at the very least, an uncomfortable feeling" (Bär, 1922). Over the years, other investigators' findings supported Millikan's hypothesis.

Conclusion: The Role and Importance of Cutting-Edge Speculative Experiments

At this stage it would be interesting to consider: What are cutting-edge speculative experiments and how do they differ from conventional experiments? According to Perl and Lee (1997): "In conventional physics experiments there is often a best way to proceed and people working in that area generally proceed the same way, with minor modifications" (p. 702). Speculative experiments become important when the scientist is groping with difficulties, future of the research cannot be predicted and stakes are high due to competing groups (peer pressure). In a recent appraisal of Millikan's methodology, Goodstein (2001) has explained cogently: "What scientist familiar with the vagaries of cutting-edge experimental work would fault Millikan for picking out [selection of oil drops] what he considered to be his most dependable measurements in order to arrive at the most accurate possible result [guiding assumption]?" (p. 57). Interestingly, in his seminal article, Holton (1978) starts by citing Medawar: "What sort of person is a scientist, and what kind of act of reasoning leads to scientific discovery.... It is no use looking to scientific 'papers', for they not merely conceal but actively misrepresent the reasoning that goes into the work they describe.... Only unstudied evidence will do – and that means listening at a keyhole" (*Art of the Soluble*, reproduced in Holton, 1978, p. 161). In the next paragraph, Holton (1978) describes what he considers to be unstudied evidence: "laboratory notebooks – first-hand documents directly rooted in the act of doing science, with all the smudges, thumbprints, and bloodstains of the personal struggle of ideas" (p. 161). Perhaps, it would not be farfetched to suggest that conventional experiments can be compared to Kuhn's (1970a) normal science and speculative (cutting-edge) experiments to revolutionary science. Furthermore, it is plausible to suggest that following Kuhn (1970b) we can consider conventional experiments to have a logic of discovery, whereas speculative experiments have a psychology of research.

The philosophical remarks of Perl and Lee (1997) in the context of the oil drop experiment (Millikan–Ehrenhaft controversy) show how the traditional scientific method as portrayed in the textbooks is not very helpful. Ehrenhaft's (having impeccable credentials) work approximated the scientific method, viz., strict adherence to experimental observations (logic of discovery) that leads to the formulations of scientific theories. Millikan, on the other hand, strictly adhered to the "hard core" (guiding assumptions, psychology of research) of his research program, which postulated the existence of the elementary electrical charge (*e*). In this context, it is interesting that Pickering (1984) depicts the role of the scientists in the traditional scientific method as passive: "In the scientists account, scientists do not appear as genuine agents. Scientists are represented rather as passive observers of nature: the facts of natural reality are revealed through experiment; experimenter's duty is simply to report what he sees; the theorist accepts such reports and supplies apparently unproblematic explanations of them" (pp. 7–8). This characterization of the scientific method approximates to a caricature of what scientists actually do and contrasts sharply with Millikan's methodology. The Millikan–Ehrenhaft controversy is a good illustration of how the traditional scientific method is perhaps found only in textbooks and not in research laboratories.

Polanyi (1964) has provided valuable insight with respect to what the scientist does and how his work is perceived as following the traditional scientific method: "A result obtained by applying strict rules mechanically, without committing anyone personally, can mean nothing to anybody. Desisting henceforth from the vain pursuit of a formalized scientific method, commitment accepts in its place the person of the scientist as the agent responsible for conducting and accrediting scientific discoveries. The scientist's procedure is of course methodical. But his methods are but the maxims of an art which he applies in his own original way to the problem of his own choice" (p. 311).

Chapter 14
Conclusion: Inductive Method as a Chimera

Acceptance of inductive generalizations as a *modus operandi*, by some scientists, philosophers of science, and textbook authors, has led Holton (1969a) to consider such understanding of progress in science as a "folkloric consensus." Some of the well-known examples of such a consensus are Einstein's special theory of relativity to have supposedly used the negative results of the Michelson–Morley experiment, falling apple in Newton's garden, and two weights dropped from the Leaning Tower of Galileo's Pisa. Although, these are well-known episodes in the history of science (some are quite folkloric indeed), presentation of physical science itself, science curriculum, and textbooks are replete with such examples. Kuhn (1977) goes to considerable length to explain the mistaken belief that quantitative data lead to new inductive generalizations of laws and theories. Furthermore, he attributes this to the prestige of modern physical science and the apparently predominant role of measurement in such research, which is bolstered by a "profoundly unhistorical source, the science text." Lakatos (1970) has referred to these interpretations of history of science as a Baconian "inductive ascent," illustrated in this monograph by historical reconstructions of the periodic table (Chapter 5), atomic theory (Chapter 6), and the covalent bond (Chapter 10). Such empiricist interpretations consider scientific progress to be dichotomous, viz., experimental observations lead to scientific laws, which later facilitate the elaboration of explanatory theories. Interestingly, the very success of modern physical science has given rise to a "myth of experimenticism" with respect to the reasons for its success based on ahistorical presentations of science textbooks. What is even more serious is the fact that most textbook authors consider inclusion of history in the textbooks as a mere chronology of events and anecdotes. Heilbron (1981) has outlined the difference between the recollections of a physicist (e.g., textbook authors) and historical reconstructions. For the former, the various intermediate steps and false trails have no significance, whereas for understanding the dynamics of scientific progress, this is what matters. This is precisely the dilemma of science textbooks and science education in general. In hindsight, all discoveries appear to be the work of geniuses who did not have to face criticisms, rivalries, conflicts, passionate debates, and alternative interpretations of data by fellow scientists. Similarly, Schwab (1974) has emphasized that scientific inquiry tends to look for patterns of change and relationships which constitute the heuristic

M. Niaz, *Critical Appraisal of Physical Science as a Human Enterprise: Dynamics of Scientific Progress*, Science and Technology Education Library 36,

(explanatory) principles of our knowledge, which have their origins not in objective facts alone, but in a conception, a deliberate construction of the mind.

As early as 1906, Duhem had critiqued the inductive method and demonstrated that "where Newton had failed, Ampère in his turn just stumbled." Both Newton and Ampère would be considered by most physicists as the founding fathers of physical science. Lakatos (1970) agreed with, endorsed, and later extended (Chapter 3) most of Duhem's historical and philosophical analyses, especially with respect to the importance of presuppositions and their role in the construction of scientific theories. In the initial phases of theory construction, presuppositions are more important rather than experimental findings. In the long run, however, verifications (based on experimental data) are more important, that is the "crown" of a theory. Both, Duhem and Lakatos agreed that in the initial phases most theories are based on inconsistent foundations and face contradictions. Despite the similarities between the philosophies of Duhem and Lakatos, in general, the latter goes beyond and provides greater understanding. For example, both were critical of the role of crucial experiments in the refutation of theories. Lakatos, however, goes beyond by suggesting that objective reasons for rejecting a research program are provided not by a crucial experiment, but rather by a rival research program. In this sense, Lakatos provides a progressive "problemshift" in our understanding of scientific progress.

James Clerk Maxwell formulated his kinetic theory of gases on the basis of a series of simplifying assumptions (presuppositions). Maxwell's presuppositions (Chapter 4) are considered to be highly speculative, involving as they do the postulation of unobserved particles exhibiting unobserved motion. Did Maxwell have an independent warrant (i.e., plausibility of the hypotheses) for his simplifying assumptions? It is plausible to suggest that Maxwell's assumptions are precisely the *ceteris paribus* clauses, which helped him to progress from simple to complex models of the gases, and can be considered as part of the *negative heuristic* (Lakatos, 1970) of his research program. Most general chemistry and physics textbooks ignore this important aspect in understanding the development of the kinetic theory of gases.

Most of the pioneering work of D. Mendeleev, with respect to the periodic table, was conducted from 1869 to 1889, before the foundations of the modern atomic theory were laid (Chapter 5). So how could Mendeleev conceptualize periodicity as a function of the atomic theory? An answer to this question precisely shows Mendeleev's ingenuity, farsightedness, creativity, and the ability to "speculate." Despite Mendeleev's own ambivalence and ambiguity, an historical reconstruction presented in this monograph does provide a convincing story of this remarkable contribution to our knowledge. Mendeleev's work also illustrates how predictions play an important role in the development of scientific theories. As Mendeleev had no empirical knowledge with respect to empty spaces in the periodic table (pertaining to missing elements), he first needed to hypothesize the existence of the missing elements based on his theoretical framework and then make predictions. Furthermore, we do not necessarily have to follow the theory/law (or for that matter ordered domain, codification scheme) dichotomy but rather it is plausible to suggest that Mendeleev's work can be considered as an "interpretative" theory, which became "explanatory" (cf. Lakatos, 1970) after the periodic law was based on atomic numbers. Most general chemistry textbooks ignore these aspects of Mendeleev's contribution.

Determination of the mass-to-charge ratio (*m*/*e*) of the cathode rays can perhaps be considered the most important experimental contribution (based on a heuristic principle) of Thomson's 1897 article. Yet, he was neither the first to do so nor the only experimental physicist.

So how did Thomson's contribution differ from that of his contemporaries? A historical reconstruction (Chapter 6) shows that Thomson's ability to speculate, elaborate alternative hypotheses and models, and perhaps most importantly formulate a theoretical framework for his experimental findings, led him to foresee and conceptualize what his contemporaries ignored. One of the important aspects of Rutherford's alpha particle experiments that deserves more attention, is that only a very small fraction (1 in 20,000) of the alpha particles deflected through large angles. Furthermore, based on the theory of probability, Rutherford showed that: (a) the chance of an alpha particle being deflected through large angles was "vanishingly small"; and (b) the probability of an alpha particle experiencing a second deflection was "negligibly small." It was precisely for these reasons that he and others found the hypothesis of single scattering (in contrast to Thomson's compound scattering), and a model of the atom with an "intense electric field," so convincing. Bohr's main objective was to explain the paradoxical stability of the Rutherford atom, and still most scientists and textbooks consider Bohr's major contribution to be the explanation of the Balmer and Paschen series of the hydrogen line spectrum. An important aspect of Bohr's model of the atom is the presence of a deep philosophical chasm: that is, in the stationary states, the atom obeys classical laws of Newtonian mechanics; on the other hand, when the atom emits radiation, it exhibits discontinuous (quantum) behavior, according to laws first proposed by Planck in 1900. Although, such contradictory grafts are fairly common in the history of science, Bohr's contemporaries failed to understand his research program and hence his theory had an adverse reception in the scientific community. Most general chemistry and physics textbooks ignore the role of heuristic principles that helped Thomson, Rutherford, and Bohr to construct their models of the atom, and also the controversies that ensued. Cooper's (1970) was the only general physics textbook that facilitated an understanding of these crucial aspects of the research methodologies of Thomson, Rutherford, and Bohr. Indeed, this textbook could provide guidelines for authors in the future.

Determination of the elementary electrical charge (*e*) has been the subject of various interpretations (Chapter 7). An important aspect of Holton's (1978) interpretation is that Millikan was not only convinced about the atomic theory of electricity, but also about the magnitude of the elementary electrical charge. According to Holton's account there were 140 experiments on an equal number of oil drops. The laboratory notebooks tell us that there were 140 drops and the published results are emphatic that there were 58 drops. What happened to the other 82 (59%) drops? Herein lies the crux of the difference in the methodologies of Ehrenhaft and Millikan. In other words, Millikan perhaps excluded drops that did not have charge equal to an integral multiple of the elementary electrical charge *e*. According to Franklin (1981), there were 175 drops in the notebooks, 68 were excluded as the apparatus was presumably not working well, another 49 were excluded even when the warm-up period was over, and of these 22 were excluded for reasons that are not very clear and of the other 27, at least 15 were excluded due to an unexpected value of *e*. Franklin generally ignored that Millikan–Ehrenhaft controversy played an important part in the determination of the elementary electrical charge, and concluded that the elimination of drops by Millikan decreased the statistical error but not the value of *e*. Hentschel (2004) has argued convincingly that Franklin seems to downplay the importance of the eliminated drops on the grounds that these could have played into the hands of Ehrenhaft. Barnes et al. (1996) have questioned Franklin's study on the grounds that it does not provide additional evidence with respect to the problematic drops as presented by Holton. Hintikka's (2005) analysis (interrogative model of inquiry) of Millikan's oil drop experiment provides an understanding that experimental data have to be validated by alternatives and that the injunction, "never to omit data" cannot be part of any absolute ethics of science. It is plausible to suggest that Ehrenhaft's methodology approximated to the traditional scientific method, which did not allow him to discard "specious drops." Millikan, on the other hand, in his publications espoused the scientific method, but in private (cf. handwritten notebooks) was fully aware of the dilemma faced and was forced to select data in order to uphold his presuppositions.

A closure to the controversy with respect to the oil drop experiment is possible if we recognize that Millikan's data selection procedure depended primarily on his perseverance with his presuppositions, viz., the existence of the elementary electrical charge and its magnitude, based on previous studies. Franklin's (1981) finding that the selection of the drops did not change the value of the elementary electrical charge (*e*), but only its statistical error carries little weight as Millikan did not perform Franklin style analyses that could have justified the exclusion of drops. Most general chemistry and physics textbooks almost completely ignore the following aspects of the determination of the elementary electrical charge: Millikan–Ehrenhaft controversy, Millikan's guiding assumptions, dependence of the elementary electrical charge on experimental variables, and Millikan's experiments as part of a progressive sequence of heuristic principles. In most traditional laboratory courses students are systematically misled into believing that the results from the oil drop experiment were straightforward and uncontroversial.

In 1905, Einstein proposed that ordinary light behaves as though it consists of a stream of independent localized units of energy that he called *light quanta* (Chapter 8). Millikan devoted considerable effort to the experimental determination of Einstein's photoelectric equation and as a consequence calculated the experimental value of Planck's constant *h* in 1916. Despite this, Millikan did not accept Einstein's theory of light quanta as late as 1924, and even considered it as "reckless." Millikan was of course, not alone in his rejection of Einstein's theory. On the contrary, most textbooks and physicists at present would consider Millikan's (1916b) experimental data as evidence for Einstein's quantum theory (hypothesis of light quanta). A plausible explanation is provided by Millikan's commitment to his presuppositions with respect to the wave theory of light, stated quite explicitly by him in many articles. The degree, to which Millikan adhered to the wave theory of light as a presupposition, is corroborated by the following scholars: Wheaton (1983), Chang (1995), and Holton (1999). Millikan's perseverance with his presuppositions helped him to determine the charge of the electron (see Chapter 7). In the case of the photoelectric effect, however, he seemed to be working with the "wrong" presuppositions. This shows clearly that presuppositions (right or wrong) form an integral part of the research program of a scientist and that at times it requires considerable amount of intellectual effort to provide grounds for the "hard core" (Lakatos, 1970) to crumble and make way for the new framework.

The purpose of the 1919 eclipse expeditions was to determine what effect, if any, is produced by a gravitational field on the path of a ray of light traversing it (Chapter 9). The predictions of both the Newton's and Einstein's values for deflection of light in the sun's gravitational field were well known in the literature before the expeditions were undertaken. Experimental evidence from the eclipse experiments was far from unequivocal, which has led to various interpretations. Some of the reasoning used by Eddington and Dyson (Dyson et al., 1920), according to which the experiments provided support for Einstein's theory, was contradictory. Let us suppose that Eddington and Dyson were not aware of Einstein's General Theory of Relativity (part of their presuppositions) and particularly of the prediction that starlight near the sun would bend. Under these circumstances experimental evidence would have been extremely uncertain, equivocal, and difficult to interpret. It is plausible to suggest that Eddington's (Dyson et al., 1920) methodology approximated the traditional scientific method, which did not allow him to discard discrepant data (this is what Millikan did, as revealed by his handwritten laboratory notebooks, see Chapter 7). Dyson and especially Eddington were fully aware as to where the theory (Einstein's) was leading them. Nevertheless, given the inductivist/positivist milieu of their scientific community they could only be guided by their "implicit" presuppositions.

It is important to note that G.N. Lewis formulated the cubic atom as early as 1902, as he wanted to explain to his chemistry students the inner structure of the atom (Chapter 10). In a letter to Robert Millikan, Lewis complained that nobody showed any interest in the model, which led him to abandon it, until 1916. The most striking and at the same time controversial feature of Lewis's theory was the formulation of the "rule of eight" or the "octet rule." Lewis postulated that the eight electrons of an octet formed the eight corners of a cube, as this provided the most stable condition for the atomic shell. Thus, the single bond was conceived of as two cubic atoms with a shared edge and the double bond as two cubes with a common face (see Figs. 10.1 and 10.2, Chapter 10). Sharing of electrons to form the covalent bond had to compete not only with the transfer of electrons (ionic bond), but was also considered "bizarre" by many chemists, as two sharing electron would produce a force of repulsion. J.J. Thomson's discovery of the electron in 1897 provided strong arguments in support of the ionic bond. Niaz (2001c) has reported that of the 27 general chemistry textbooks (published in the USA.) analyzed, only 3 described satisfactorily the role played by Lewis's cubic atom. One of the textbooks mentioned and none described satisfactorily that Lewis's idea of sharing electrons had to compete with the transfer of electrons. Only two textbooks explained that Lewis's covalent bond was not induced from experimental evidence, but derived from the cubic atom. A historical reconstruction shows that the Lewis's cubic atom was the first attempt to explain the stability of the covalent bond and later the quantum theory provided further support when Pauli introduced his exclusion principle. Transition from Lewis's cubic atom → Pauli's exclusion principle → what next provides an illustration of how scientific knowledge is tentative. Eight textbooks explained satisfactorily that the Pauli's exclusion principle provided an explanation of the covalent bond.

Both physicists and philosophers of science now agree that the Copenhagen interpretation is not the only possible consistent theory of quantum mechanics (Chapter 11). Alternative points of view are considered as perfectly consistent, including Bohm's interpretation of "hidden variables." According to Cushing (1991), the question is whether we are capable of truly *understanding* quantum phenomena, as opposed to simply *accepting* the formalism. Similarly, Styer (2000) has emphasized the need to delve deeply into quantum mechanics as a set of physical ideas rather than as an elaborate and somewhat mystical algorithm for solving problems in atomic physics. Kuhn (1978) concluded that it was Paul Ehrenfest and Albert Einstein who first recognized that the black-body law could not be derived without restricting resonator energy to integral multiples of $h\nu$. In other words, Planck in 1900 simply introduced an approximate mathematical quantization for convenience in doing the calculations. In Cushing's (1991) scheme this would constitute a simple algorithm for reproducing observed data. On the other hand, the physical significance of the quantum hypothesis was first explained by Einstein. Despite skepticism on the part of some historians with respect to Kuhn's controversial thesis, Brush (2000) has shown that a historical reconstruction provides a clear confirmation of Kuhn's thesis.

According to the standard Copenhagen interpretation, one cannot discount the effect of observations so as to frame an observer-independent picture of the world. In contrast, according to the Bohmian interpretation for velocity, the unmeasured particle is actually standing still! Measurement disturbs the situation, freeing the wave function, which guides the particle into motion (Fine 1996a). At this stage, one wonders: Why did the Copenhagen physicists hold fast to the notions of observer dependence and inevitable acausality? Bohm (1957) has explained in cogent terms that this can be attributed to the widespread popularity of positivism among physicists during the early twentieth century, according to which we should not postulate the existence of entities, which cannot be observed by methods that are already available. Bohm's historical reconstruction is indeed illuminating and indicates quite clearly as to how in the history of science the quest for further and deeper understanding through as yet unknown "hidden" variables has been a source of controversy and conflict. Niaz and Fernández (2008) have reported that none of the general chemistry textbooks referred to: (a) the difference between Planck's and Einstein's contributions with respect to the origin of the quantum hypothesis; and (b) Bohm's interpretation of quantum mechanics. According to Bowman (2002), this understanding based on Bohmian mechanics, in turn, can help develop physical intuition – an elusive commodity in the quantum world.

De Broglie (1924) in a seminal paper explored the reconciliation of light quanta (used by Einstein to explain the photoelectric effect) with the wave theory, which in turn was based on strong experimental evidence (Chapter 12). Millikan and many other physicists conceded that as the photoelectric effect could not be explained by the classical wave theory, it may at the most need some modifications and not its rejection. Maurice de Broglie had suggested that the theory of wave–particle duality should also include an experimental part. De Broglie declined, saying he was not an experimentalist. These days, however, we have no such inhibitions about accepting the quark model, even though no one has ever isolated a quark (Perl & Lee, 1997). This facilitates an understanding of how scientific research methodology has evolved over the last 7 decades. Davisson and Germer (1927) later found that not only experiments were difficult to perform, but also their interpretation was elusive. Acceptance of Louis de Broglie's ideas was difficult due to his previous controversies with two influential schools of physicists (Copenhagen's and Munich's). Einstein's early support was crucial, in later convincing Schrödinger of de Broglie's thesis. Conceptually, Schrödinger aligned himself with Einstein in his controversy with the Copenhagen school of quantum mechanics on important issues such as causality, probability, and determinism. De Broglie's difficulties were compounded by the "positivist milieu" of the Copenhagen school on the one hand, and the theory itself was not fully understood and to make matters worse, ill-defended (Bonk, 1994). This historical reconstruction is a good indicator of how prior epistemological beliefs of the scientists play an important role in the acceptance of new ideas.

Although, theorists have become convinced that free quarks do not exist, but are confined within hadrons of which they are parts, various research groups (including that of Martin Perl at Stanford) continue their efforts to isolate quarks (Chapter 13). Perl and colleagues have used an automated Millikan liquid drop method with improvements based on modern technology. Given the difficulties involved in cutting-edge experimental work, Perl and Lee (1997) have proposed a philosophy of speculative experiments. In order to justify their methodology, these authors have made eight philosophical remarks. It appears that these remarks can be sustained by referring to the research methodology of Millikan in the context of the oil drop experiment, and also facilitate an understanding of the Millikan–Ehrenhaft controversy. Of the eight philosophical remarks the following are more pertinent: (a) in sensible, speculative experimental work, the research must extend beyond established experimental knowledge; (b) choices in the design of speculative experiments usually cannot be made simply on the basis of pure reason; (c) a beautiful experiment is always pleasant; and (d) in a speculative research it is wise to minimize experimental restrictions. A reconstruction of Millikan's experimental methodology (see Chapter 7) shows that he used similar speculative guidelines in his data reduction procedures. However, there is an essential difference between the philosophies of Millikan and Perl. What Perl has acknowledged publicly, Millikan somehow tried to "conceal." Had it not been for the painstaking efforts of Holton (1978) in scrutinizing his original handwritten notebooks, history of science would have recognized Millikan's methodology as yet another case in which experimental data led to unambiguous theoretical formulations. In this context, Martin Perl's philosophy of speculative experiments has made an important contribution to our understanding of history and philosophy of science.

The Inductive Method

According to Duhem (1914): "The teaching of physics by the purely inductive method such as Newton defined it is a chimera. Whoever claims to grasp this mirage is deluding himself and deluding his pupils. He is giving them, as facts seen, facts merely foreseen; as precise observations, rough reports; as performable procedures, merely ideal experiments; as experimental laws, propositions whose terms cannot be taken as real without contradictions. The physics he expounds is false and falsified" (p. 205). These are thought-provoking ideas that provide us an agenda for understanding and teaching science, presented almost a hundred years ago. Actually, Duhem (1914, pp. 268–270) explicitly endorses an historical method for teaching science, which has the following advantages: (a) retracing the long series of errors and hesitations (false trails) preceding the discovery of each principle; (b) puts the student on guard against false evidence; (c) recalling the vicissitudes of the contending schools; (d) exhuming doctrines once triumphant from the oblivion in which they lie; and (e) even the most attractive theories are only provisional representations and not definitive explanations. Similarly, Lakatos (1999) argued with respect to the difficulties involved in proving a scientific proposition from facts. Holton (1969a) has expressed the same point cogently, by recognizing that science textbooks place a high value on clear, unambiguous, inductive reasoning, which facilitate a clear genetic link from experiment to theory. Furthermore, Holton goes on to emphasize that such an approach is to some degree distorted, and based on "*an outdated or erroneous philosophy of science*." The outdated and erroneous philosophy of science that Holton refers to is, of course the inductive method. Similarly, Crowe (1999) has endorsed the same message in the following terms: "If pressed to specify one message about science that needs to be conveyed to liberal arts students, it is that 'naive inductivism' is not, nor could it have been, the method of science" (p. 63). Cooper has suggested that many people believe that science is actually done that way, that is the traditional method in which more technique can be crammed into a given time, at the expense of ideas (reproduced in Niaz et al., 2009). If we want our students to be important scientists, then ideas within an historical context are more important (reproduced in Niaz et al., 2009).

If we tell students that "science is empirical," we shall be denying them an important aspect of the nature of science. For example, Kaufmann, Wiechert, and Thomson (see Chapter 6) determined experimentally the mass-to-charge ratio of cathode rays at about the same time and their values agreed to a fair degree. If science had been entirely empirical, the experimental work of all three would have the same importance. The crucial difference between these three scientists was that only Thomson ventured to hypothesize (in part through speculations) based on a heuristic principle that cathode rays were universal charged particles and thus constituents of all atoms.

According to Rigden and Stuewer (2005), nowhere is the contrast between quantitative and qualitative more apparent than in teaching physics to students. In the classroom, the quantitative dominates and students respond accordingly, that is learn (basically memorize) algorithms to solve quantitative problems to obtain the correct answer. The paradox is that solving quantitative problems based on the inductive method do not facilitate an understanding of the subject. Research in science education has provided considerable evidence with respect to how problem solving based on memorization of algorithms does not facilitate conceptual understanding (Niaz, 1995; Niaz & Robinson, 1992; Pickering, 1990).

Milieu of the Time

Cooper (1992) has emphasized that if we want students to understand why certain questions are asked at a particular time and why experiments are performed, then it is important to provide them with the *milieu of the time*, based on what the scientific community was thinking (viz., its presuppositions), and this requires an historical perspective. Interestingly, according to Cooper, in Copernicus' time the null result of the Michelson–Morley experiment would have been taken to confirm the belief that the earth was at rest at the center of the universe. In other words, if we want our students to understand as to how and why an experiment was done, it would be helpful to retrace the false trails. More recently, Cooper has reiterated this point that "presenting physics in historic context makes it more understandable" (reproduced in Niaz et al., 2009).

Do Varying Interpretations of Experimental Data Undermine the Objective Nature of Science?

According to Cooper, varying interpretations of the same experimental data do not undermine the objective nature of science, but rather deepen our understanding (reproduced in Niaz et al., 2009). Would Newton, Ampère, Thomson, Rutherford, Bohr, Millikan, Ehrenhaft, and Eddington have agreed with this statement? It is plausible to conclude that they would have denied it and in this very sense there is a progressive problemshift in our understanding of scientific research methodology. Cooper goes beyond by suggesting that the "vision of the scientist" can be compared to that of a painter. For example, the impressionists were accused of not being able to see things as they are. However, having imposed their vision of the world, it has become a cliché now to see things as the impressionists did. Interpretation of data from the 1919 eclipse experiments was based on "Eddington's vision" (which was not entirely justified then, Chapter 9) and later the scientific community accepted that vision.

Role of Presuppositions of the Scientist

In order to pursue further possible changes in research methodology, Niaz et al. (2009) asked Leon Cooper the following question: "In your opinion, how closely is the conception of an experiment tied to the presuppositions of the scientists?" Cooper's response is an eye-opener and worth reproducing at length:

> Experiments are designed to try to answer questions that seem important at the time. Sometimes there is general agreement about what we'd like to know. For example, there are many things we would like to know that relate to current string theories, but nobody can build machines to provide the means of testing them. One of the values of a theoretical structure is that it focuses the questions that are relevant. There are so many possibilities, one could poke around endlessly. The experimentalist wants to have some reason for spending the time – and often it's an enormous amount of time – to do difficult experiments. The decision is usually based on what is possible and, from the current point of view (*data, theory and conjecture*) what is important. (Reproduced in Niaz et al., 2008, emphasis added)

In order to understand the dilemma faced by the scientist, with respect to research methodology, Cooper refers to string theories. A recent study has evaluated string theory from a history and philosophy of science perspective and concluded that:

> [T]he fact that string theory has not been corroborated by any direct empirical evidence thus far seems to render it a mere theoretical speculation....For many years now, the string community has been one of the largest communities in all of theoretical physics and has produced the majority of the field's top-cited papers....The fact that an entirely unconfirmed speculative idea can assume such a prominent position in a mature scientific field is quite astonishing. (Dawid, 2006, p. 299)

The underlying rationale of Cooper and Dawid may be the same, viz., experiments provide invaluable information about the real world, which is difficult to understand; however, experiments are difficult to perform and their meaning is elusive. This dilemma paves the way for a scientist to integrate data, theory, and conjecture and, in doing so, legitimizes the role played by presuppositions. This monograph has highlighted the role of presuppositions in the work of various scientists: Newton and Ampère (Chapter 2); Maxwell (Chapter 4); Mendeleev (Chapter 5); Thomson, Rutherford, and Bohr (Chapter 6); Millikan (Chapter 7); Lewis (Chapter 10); Eddington (Chapter 9); De Broglie (Chapter 12); and Perl (Chapter 13).

Experiments and Conjectures

To pursue further, Niaz et al. (2009) asked Leon Cooper another question: "In your opinion, what is the relationship between *experiments* and *conjectures*? Are they always interdependent?" Once again, Cooper's response reflects the dynamics of scientific progress and we have reproduced at length:

> [A]t a given time there is a view of the world. Consider molecular biology or genetics today. This view of the world is based on actual data, on people's beliefs as to what the data means, and on conjectures about data that isn't presently available. *So the current view is almost always a mixture of data, hypotheses, theoretical ideas, and conjectures*. In designing an experiment one often tries to sort among the various conjectures or test a theoretical postulate or prediction. This is the process that moves the entire field forward. (Reproduced in Niaz et al., 2009, emphasis added)

At this stage it would be interesting to know if teachers, developers of science textbooks, and curricula would accept Cooper's advice that the current view in a given science (e.g., molecular biology) "is almost always a mixture of data, hypotheses, theoretical ideas and conjectures." This provides an important guideline not only for science education, but also in understanding "science in the making" and consequently scientific research methodology.

Data from Experiments Do Not Unambiguously Lead to the Formulation of Theories

Similar experimental data can be interpreted in various ways. Some of the examples discussed in this monograph illustrate this aspect of scientific research methodology cogently: (a) Rutherford's hypothesis of single scattering and Thomson's hypothesis of compound scattering to explain alpha particle experiments (Chapter 6); (b) Millikan's hypothesis of elementary electrical charge (electron) and Ehrenhaft's hypothesis of fractional charges (sub-electron), to explain very similar oil/liquid drop experiments (Chapter 7); (c) data from bending of light in the 1919 eclipse experiments could be interpreted to provide support for Newton's or Einstein's theories, or even perhaps for none of the two theories (Chapter 9); (d) Millikan's experimental data for the determination of Planck's constant (*h*) did not provide evidence for Einstein's theory of light quanta (Chapter 8); and (e) Bohm's theory of "hidden variables" could explain almost all the experimental phenomena explained by the Copenhagen interpretation of quantum mechanics (Chapter 11).

Role of Inconsistencies

Cooper has also provided important insight with respect to the role of inconsistencies in the construction of scientific theories:

> Scientific theories, when they're finished, are supposed to be consistent. But in their construction they often contain inconsistencies; these inconsistencies are among the problems that people try to resolve. For example, quantum electrodynamics, at least as viewed in the 1950s, was believed to be inconsistent – problems related to the infinities and so on. On the other hand, there was no doubt that some components of the theory, for example, theoretical calculations and experimental measurements of the magnetic moment of the electron were so marvellously in agreement, that this structure would be retained in any new theory. But the inconsistencies might lead to new theoretical ideas. This is exactly what happened in the case of the Bohr atom; there were obvious inconsistencies, but the Bohr atom was so successful that the fruitful approach was to find out how to remove the inconsistencies and still retain the structure of the theory. (Reproduced in Niaz et al., 2009)

This monograph has provided examples of various episodes in which scientific theories were based on inconsistent foundations: Maxwell's kinetic theory (Chapter 4); Mendeleev's periodic table (Chapter 5), Bohr's model of the atom (Chapter 6), and Einstein's hypothesis of light quanta (Chapter 8).

Role of Speculations

It is interesting to observe that of all the historical episodes (except one) discussed in this monograph, none of the scientists alluded to and much less refer to the role played by speculations. Does this mean that speculation has no place in the development of scientific theories? The only exception was Martin Perl, who developed a philosophy of speculative experiments to isolate quarks (Perl & Lee, 1997). In Chapter 13, we also presented the views of Leon Cooper with respect to Perl's philosophy. Cooper endorsed Perl's philosophy almost entirely. This makes interesting reading, as Leon Cooper in a sense goes beyond Perl and Lee (1997), by emphasizing not only speculations, but also intuition guided by facts and conjectures.

Role of Controversies

Controversies play an important part in scientific research and are a common denominator in all the historical episodes discussed in this monograph. Machamer et al. (2000) have argued that although nobody would deny that science in the making has been replete with controversies, the same people often portray science as "the uncontroversial rational human endeavor par excellence." Again, it is no surprise that in most parts of the world, both science textbooks and curricula either ignore or do not emphasize the importance of controversies in the development of scientific knowledge.

Progress in Scientific Research Methodology

How scientific research methodology seems to have progressed, if we compare the following sequence of philosophers of science: Duhem, Polanyi, Holton, Lakatos, and Giere? It is even more instructive to compare the methodologies of these philosophers of science with those of Newton, Ampère, Maxwell, Thomson, Rutherford, Bohr, Einstein, Millikan, Eddington, De Broglie, Bohm, Perl, and Cooper, leading physicists of their respective periods, ranging over a period of almost 300 years. It is plausible to suggest that starting from Perl and Cooper (although Einstein, De Broglie, and Bohm came quite close), scientists now understand scientific research methodology more in tune with developments in philosophy of science. Considering the fact that these philosophers of science were originally trained in chemistry, mathematics, and physics, makes an interesting mosaic of thinking. Indeed both Duhem and Lakatos would have enjoyed to contemplate how they prepared the terrain for Perl and Cooper.

References

Abd-El-Khalick, F. & Lederman, N.G. (2000). Improving science teachers' conceptions of nature of science: A critical review of the literature. *International Journal of Science Education, 22*, 665–701.

Abd-El-Khalick, F. (2004). Over and over again: College students' views of nature of science. In L.B. Flick & N.G. Lederman (eds), *Scientific Inquiry and Nature of Science: Implications for Teaching, Learning, and Teacher Education* (pp. 389–425). Dordrecht, The Netherlands: Kluwer.

Abd-El-Khalick, F. (2005). Developing deeper understanding of nature of science: The impact of a philosophy of science course on preservice science teachers' views and instructional planning. *International Journal of Science Education, 27*, 15–42.

Abd-El-Khalick, F. & Akerson, V.L. (2004). Learning about nature of science as conceptual change: Factors that mediate the development of preservice elementary teachers' views of nature of science. *Science Education, 88*, 785–810.

Abd-El-Khalick, F., Bell, R.L. & Lederman, N.G. (1998). The nature of science and instructional practice: Making the unnatural natural. *Science Education, 82*, 417–436.

Abd-El-Khalick, F., Waters, M. & Le, A. (2008). Representation of nature of science in high school chemistry textbooks over the past decades. *Journal of Research in Science Teaching, 45*(7), 835–855.

Abegg, R. (1904). Die valenz und das periodische system. *Zeitschrift fur Anorganische Chemie, 39*, 330–380.

Achinstein, P. (1987). Scientific discovery and Maxwell's kinetic theory. *Philosophy of Science, 54*, 409–434.

Achinstein, P. (1991). *Particles and Waves: Historical Essays in the Philosophy of Science.* New York: Oxford University Press.

Akerson, V.L., Morrison, J.A. & McDuffie, A.R. (2006). One course is not enough: Preservice elementary teachers' retention of improved views of nature of science. *Journal of Research in Science Teaching, 43*, 194–213.

Allchin, D. (2005). Letters. *Science, 308*(3 June), 1409–1410.

Andrews, T. (1869). On the continuity of the gaseous and liquid states of matter. (First Bakerian Lecture, 1869. Reproduced in *Scientific Papers*, pp. 296–317, London, 1896).

Ardac, D. (2002). Solving quantum number problems: An examination of novice performance in terms of conceptual based requirements. *Journal of Chemical Education, 79*, 510–513.

Arriassecq, I. & Greca, I.M. (2007). Approaches to the teaching of special relativity theory in high school and university textbooks. *Science & Education, 16*(1), 65–86.

Arsem, W.C. (1914). A theory of valence and molecular structure. *Journal of American Chemical Society, 36*, 1655–1675.

Babbage, C. (1830). *Reflections on the Decline of Science in England and on Some of Its Causes.* London: B. Fellowes.

Bachelard, G. (1949). The philosophical dialectic of the concepts of relativity. In P.A. Schilpp (ed.), *Albert Einstein: Philosopher-Scientist* (pp. 566–568). Evanston, IL: Library of Living Philosophers.

Balmer, J.J. (1885). *Annalen der Physik und Chemie*, 25, 80–85.
Barnes, B., Bloor, D. & Henry, J. (1996). *Scientific Knowledge: A Sociological Analysis*. Chicago, IL: University of Chicago Press.
Bär, R. (1922). *Naturwissenschaften*, *10*, 344–345.
Bell, R., Abd-El-Khalick, F., Lederman, N.G., McComas, W.F. & Matthews, M.R. (2001). The nature of science and science education: A bibliography. *Science & Education*, *10*, 187–204.
Beller, M. (1999). *Quantum Dialogue: The Making of a Revolution*. Chicago, IL: University of Chicago Press.
Bensaude-Vincent, B. (1986). Mendeleev's periodic system of chemical elements. *British Journal for the History of Science*, *19*, 3–17.
Bianchini, J.A. & Colburn, A. (2000). Teaching the nature of science through inquiry to prospective elementary teachers: A tale of two researchers. *Journal of Research in Science Teaching*, *37*, 177–209.
Blanco, R. & Niaz, M. (1997). Epistemological beliefs of students and teachers about the nature of science: From "Baconian inductive ascent" to the "irrelevance" of scientific laws. *Instructional Science*, *25*, 203–231.
Blanco, R. & Niaz, M. (1998). Baroque tower on a Gothic base: A Lakatosian reconstruction of students' and teachers' understanding of structure of the atom. *Science & Education*, *7*, 327–360.
Bodner, G.M. & Pardue, H.L. (1989). *Chemistry: An Experimental Science*. New York: Wiley.
Bohm, D. (1952). A suggested interpretation of the quantum theory in terms of "hidden" variables, I and II. *Physical Review*, *85*, 166–179/180–193.
Bohm, D. (1957). *Causality and Chance in Modern Physics*. London: Routledge & Kegan Paul.
Bohm, D. (1976). On the creation of a deeper insight into what may underlie physical law. In M.M. Flato et al. (eds), *Quantum Mechanics, Determinism, Causality and Particles* (pp. 4–5). Dordrecht, The Netherlands: Reidel.
Bohm, D. (1980). *Wholeness and the Implicate Order*. London: Routledge & Kegan Paul.
Bohm, D. & Vigier, J.-P. (1954). Model of the causal interpretation of quantum theory in terms of a fluid with irregular fluctuations. *Physical Review*, *96*, 208–216.
Bohr, N. (1913a). On the constitution of atoms and molecules, Part I. *Philosophical Magazine*, *26*(Series 6), 1–25.
Bohr, N. (1913b). On the constitution of atoms and molecules, Part II: Systems containing only a single nucleus. *Philosophical Magazine*, *26*(Series 6), 476–502.
Bohr, N. (1913c). On the constitution of atoms and molecules, Part III: Systems containing several nuclei. *Philosophical Magazine*, *26*(Series 6), 857–875.
Bohr, N. (1913–1914). Atomic models and X-ray spectra. *Nature*, *92*, 553.
Boltzmann, L. (1868). Studien uber das gleichgewicht der lebendigen kraft zwischen bewegten materiellen Punkten. *Wissenschaftliche Abhandlungen*, *1*, 49–96.
Boltzmann, L. (1871). Analytischer beweis des zweiten hauptsatzes der mechanischen wärmetheorie aus den sätzen uber gleichgewicht der lebendigen kraft. *Wissenschaftliche Abhandlungen*, *1*, 288–308.
Boltzmann, L. (1872). Weitere studien uber das wärmegleichgewicht unter gasmolekulen. *Wissenschaftliche Abhandlungen*, *1*, 316–402.
Bonk, T. (1994). Why has de Broglie's theory been rejected? *Studies in History and Philosophy of Science*, *25*(3), 375–396.
Boorse, H.A. & Motz, L. (1966). *The World of the Atom*. New York: Basic Books.
Bowman, G.E. (2002). Bohmian mechanics as a heuristic device: Wave packets in the harmonic oscillator. *American Journal of Physics*, *70*, 313–318.
Bray, W.C. & Branch, G.E.K. (1913). Valence and tautomerism. *Journal of American Chemical Society*, *35*, 1440–1447.
Brady, J.E., Russell, J.W. & Holum, J.R. (2000). *Chemistry: Matter and Its Changes*. New York: Wiley.
Brito, A., Rodríguez, M.A. & Niaz, M. (2005). A reconstruction of development of the periodic table based on history and philosophy of science and its implications for general chemistry textbooks. *Journal of Research in Science Teaching*, *42*, 84–111.

Brown, T.L. & LeMay, H.E. (1988). *Chemistry: The Central Science* (7th ed., Spanish). Englewood Cliffs, NJ: Prentice-Hall.

Brush, S.G. (1974). Should the history of science be rated X. *Science, 183*, 1164–1172.

Brush, S.G. (1976). *The Kind of Motion We Call Heat: A History of the Kinetic Theory of Gases in the 19th Century*. New York: North-Holland.

Brush, S.G. (1978). Why chemistry needs history – and how it can get some. *Journal of College Science Teaching, 7*, 288–291.

Brush, S.G. (1980). The chimerical cat: Philosophy of quantum mechanics in historical perspective. *Social Studies of Science, 10*, 393–447.

Brush, S.G. (1989). Prediction and theory evaluation: The case of light bending. *Science, 246*, 1124–1129.

Brush, S.G. (1996). The reception of Mendeleev's periodic law in America and Britain. *Isis, 87*, 595–628.

Brush, S.G. (1999). Why was relativity accepted? *Physics in Perspective, 1*(2), 184–214.

Brush, S.G. (2000). Thomas Kuhn as a historian of science. *Science & Education, 9*, 39–58.

Brush, S.G. (2005). Letters. *Science, 308*(3 June), 1410.

Brush, S.G. (2007a). Predictivism and the periodic table. *Studies in History and Philosophy of Science, 38*, 256–259.

Brush, S.G. (2007b). How ideas became knowledge: The light quantum hypothesis 1905–1935. *Historical Studies in the Physical Sciences, 37*(2), 205–246.

Burbules, N.C. & Linn, M.C. (1991). Science education and philosophy of science: Congruence or contradiction? *International Journal of Science Education, 13*, 227–241.

Burns, R.A. (1995). *Fundamentals of Chemistry* (2nd ed., Spanish). Englewood Cliffs, NJ: Prentice-Hall.

Campbell, D.T. (1988). The experimenting society. In E.S. Overman (ed.), *Methodology and Epistemology for Social Science* (pp. 290–314). Chicago, IL: University of Chicago Press (first published 1971).

Cartwright, N. (1983). *How the Laws of Physics Lie*. Oxford: Clarendon Press.

Cartwright, N. (1989). *Nature's Capacities and Their Measurement*. Oxford: Clarendon Press.

Cartwright, N., Cat, J., Fleck, L. & Uebel, T.E. (1996). *Otto Neurath: Philosophy Between Science and Politics*. Cambridge: Cambridge University Press.

Chang, H. (1995). The quantum counter-revolution: Internal conflicts in scientific change. *Studies in History and Philosophy of Science, 26*, 121–136.

Chang, R. (1981) *Chemistry*. New York: Random House.

Chiappetta, E.L., Sethna, G.H. & Dillman, D.A. (1991). A quantitative analysis of high school chemistry textbooks for scientific literacy themes and expository learning aids. *Journal of Research in Science Teaching, 28*, 939–951.

Christie, M. (1994). Philosophers versus chemists concerning "laws of nature." *Studies in History and Philosophy of Science, 25*, 613–629.

Christie, J.R. & Christie, M. (2003). Chemical laws and theories: A response to Vihalemm. *Foundations of Chemistry, 5*, 165–174.

Clark, P. (1976). Atomism versus thermodynamics. In C. Howson (ed.), *Method and Appraisal in the Physical Sciences: The Critical Background to Modern Science, 1800–1905* (pp. 41–105). Cambridge: Cambridge University Press.

Clausius, R. (1857). On the nature of the motion which we call heat. *Philosophical Magazine, 14*, 108–127.

Clough, M.P. (2006). Learners' responses to the demands of conceptual change: Considerations for effective nature of science instruction. *Science & Education*, 15, 463–494.

Collingwood, R.G. (1964). *The Idea of Nature*. Oxford: Clarendon Press.

Collingwood, R.G. (1966). *An Essay on Metaphysics*. Oxford: Clarendon Press (first published 1940).

Collins, H. & Pinch, T. (1993). *The Golem: What Everyone Should Know About Science*. Cambridge: Cambridge University Press.

Collins, H. & Pinch, T. (1998). *The Golem: What You Should Know About Science* (2nd ed.). Cambridge: Cambridge University Press.

Cooper, L.N. (1970). *An Introduction to the Meaning and Structure of Physics* (short edition). New York: Harper & Row.
Cooper, L.N. (1992). *Physics: Structure and Meaning*. Hanover, NH: University Press of New England.
Crease, R.M. (2002). Critical point: The most beautiful experiment. *Physics World, 15*(9), 19–20.
Crease, R.M. (2003). *The Prism and the Pendulum: The Ten Most Beautiful Experiments in Science*. New York: Random House.
Crookes, W. (1877). The chemistry of the future. *Quarterly Journal of Science*, N.S., *7*, 289–306.
Crookes, W. (1879). On radiant matter. *Nature*, *20*, 419.
Cropper, W.H. (1970). *The Quantum Physicists*. New York: Oxford University Press.
Crowe, M.J. (1990). Duhem and history and philosophy of mathematics. *Synthese*, *83*, 431–447.
Crowe, M.J. (1999). Pierre Duhem, the history and philosophy of physics, and the teaching of physics. *Physics in Perspective*, *1*, 54–64.
Crowther, J.G. (1910). *Proceedings of the Royal Society* (vol. lxxxiv). London: Royal Society.
Crowther, J.G. (1914). *Molecular Physics*. Philadelphia, PA: P. Blackiston.
Crowther, J.G. (1974). *The Cavendish Laboratory 1874–1974*. New York: Science History Publications.
Curie, M. (1914). Les radio-éléments et leur classification. In *Les progrès de la physique moléculaire*. Paris: Gauthier-Villars.
Cushing, J.T. (1989). The justification and selection of scientific theories. *Synthese*, *78*, 1–24.
Cushing, J.T. (1991). Quantum theory and explanatory discourse: Endgame for understanding. *Philosophy of Science*, *58*, 337–358.
Cushing, J.T. (1995). Quantum mechanics, underdetermination and hermeneutics. *Science & Education*, *4*(2), 137–146.
Cushing, J.T. (1996). The causal quantum theory program. In J.T. Cushing, A. Fine & S. Goldstein (eds), *Bohmian Mechanics and Quantum Theory: An Appraisal* (pp. 1–19). Dordrecht, The Netherlands: Kluwer.
Darling, K.M. (2002). The complete Duhemian underdetermination argument: Scientific language and practice. *Studies in History and Philosophy of Science*, *33*(3), 511–533.
Darrigol, O. (1993). Strangeness and soundness in Louis de Broglie's early works. *Physis: Rivista Internazionale de Storia della Scienza*, *30*(2–3), 303–372.
Darwin, C.G. (1962). Address to the Rutherford Jubilee Conference, Manchester.
Davisson, C. & Germer, L.H. (1927a). Diffraction of electrons by a crystal of Nickel. *Physical Review*, *30*(6), 705–740.
Davisson, C. & Germer, L.H. (1927b). The scattering of electrons by a single crystal of nickel. *Nature*, *119*, 558.
Davisson, C. & Kunsman, C.H. (1923). The scattering of low speed electrons by. platinum and magnesium. *Physical Review*, *22*, 242–258.
De Broglie, L. (1909). *Le Radium*, *6*, 203.
De Broglie, L. (1922). *Journal de Physique*, *3* (Series VI), 422.
De Broglie, L. (1923a). Ondes et quanta. *Comptes Rendus*, *177*, 507–510, 548–550, 630–632.
De Broglie, L. (1923b). Waves and quanta. *Nature*, *112*, 540.
De Broglie, L. (1924). A tentative theory of light quanta. *Philosophical Magazine*, *47*(Series 6), 446–458.
De Broglie, L. (1953). Foreword. In P. Duhem (1914), *The Aim and Structure of Physical Theory*. New York: Atheneum.
De Milt, C. (1951). The congress at Karlsruhe. *Journal of Chemical Education*, *28*, 421–425.
De Regt, H.W. (1996). Philosophy and the kinetic theory of gases. *British Journal for the Philosophy of Science*, *47*, 31–62.
Dickerson, R.E., Gray, H.B., Darensbourg, M.Y. & Darensbourg, D.J. (1984). *Chemical Principles* (4th ed.) Menlo Park, CA: Benjamin Cummings.
Dirac, P.A.M. (1977). Ehrenhaft, the subelectron and the quark. In C. Weiner (ed.), *History of Twentieth Century Physics* (pp. 290–293). New York: Academic.

Dmitriev, I.S. (2004). Scientific discovery *in statu nascendi*: The case of Dmitrii Mendeleev's periodic law. *Historical Studies in the Physical Sciences*, *34*, 233–275.

Dobson, K., Lawrence, I. & Britton, P. (2000). The A to B of quantum physics. *Physics Education*, *35*, 400–405.

Donovan, A., Laudan, L. & Laudan, R. (1988). *Scrutinizing Science: Empirical Studies of Scientific Change*. Dordrecht, The Netherlands: Kluwer.

Dorling, J. (1970). Maxwell's attempt to arrive at non-speculative foundations for the kinetic theory. *Studies in History and Philosophy of Science*, *1*, 229–248.

Duhem, P. (1914). *The Aim and Structure of Physical Theory* (2nd ed., trans. Philip P. Wiener, first published 1906 as *La théorie physique: Son objet, sa structure* by Marcel Rivière Cie, Paris. First English translation by Princeton University Press, 1954 and later by Atheneum in 1962. The book was originally published as a series of articles in French in the years 1904–05). New York: Atheneum.

Duhem, P. (1962). *The aim and structure of physical theory* (trans. P.P. Wiener). New York: Atheneum (original work published in 1906).

Dyson, F.W., Eddington, A.S. & Davidson, C. (1920). A determination of the deflection of light by the sun's gravitational field, from observations made at the total eclipse of May 29, 1919. *Royal Society Philosophical Transactions*, *220*, 291–333.

Earman, J. & Glymour, C. (1980). Relativity and eclipses: The British eclipse expeditions of 1919 and their predecessors. *Historical Studies in the Physical Sciences*, *11*(1), 49–85.

Ehrenhaft, F. (1909). Eine Methode zur Bestimmung des Elektrischen Elementarquantums. *Physikalische Zeitschrift*, *10*, 308–310.

Einstein, A. (1905). über einen Erzeugung und Verwandlung des Lichtes betreffenden heuristischen Gesichtspunkt. *Annalen der Physik*, *17*, 132–148.

Einstein, A. (1909a). Zum gegenwärtigen stand des strahlungsproblems. *Physikalische Zeitschrift*, *10*, 185–193.

Einstein, A. (1909b). Über die Entwicklung unsere anschauungen über das wesen und die konstitution der strahlung. *Physikalische Zeitschrift*, *10*, 817–825.

Einstein, A. (1916). Zur Quantentheorie der Strahlung. *Zürich Mitteilungen*, *18*, 47–62.

Einstein, A. (1925). Quantentheorie des einatomigen idealen gases. 2. Abhandlung. *Preuss. Akad. d. Wiss., Sitzingsber*, *1*, 3–14.

Eisberg, R.M. (1973). *Fundamentals of Modern Physics*. New York: Wiley.

Elkana, Y. (1974). *The Interaction Between Science and Philosophy*. Atlantic Highlands, NJ: Humanities Press.

Ellis, B.D. (1991). Idealization in science. In C. Dilworth (ed.), *Idealization IV: Intelligibility in Science*. Amsterdam: Rodopi.

Elsasser, W. (1925). *Naturwissenschaften*, *13*, 711.

Ernest, P. (1994). The dialogical nature of mathematics. In P. Ernest (ed.), *Mathematics, Education, and Philosophy: An International Perspective* (pp. 33–48). London: Falmer Press.

Fairbank, W.M. & Franklin, A. (1982). Did Millikan observe fractional charges on oil drops? *American Journal of Physics*, *50*, 394–397.

Falconer, I. (1987). Corpuscles, electrons, and cathode rays: J.J. Thomson and the "discovery of the electron." *British Journal for the History of Science*, *20*, 241–276.

Feyerabend, P.K. (1970). Against method: Outline of an anarchist theory of knowledge. In M. Radner & S. Winokur (eds), *Minnesota Studies in the Philosophy of Science* (vol. IV, pp. 17–130). Minneapolis, MN: University of Minnesota Press.

Feynman, R. (1967). *The Character of Physical Law*. Cambridge, MA: MIT Press.

Feynman, R. (1985). *The Strange Theory of Light and Matter*. London: Penguin.

Fine, A. (1996a). On the interpretation of Bohmian mechanics. In J.T. Cushing, A. Fine & S. Goldstein (eds), *Bohmian Mechanics and Quantum Theory: An Appraisal*. Dordrecht, The Netherlands: Kluwer.

Fine, A. (1996b). *The Shaky Game: Einstein, Realism and the Quantum Theory* (2nd ed). Chicago, IL: University of Chicago Press.

FitzGerald, G. (1897). Dissociation of atoms. *The Electrician*, *39*, 103.

Feynman, R. (1967). *The Character of Physical Law*. Cambridge, MA: MIT Press.
Frankel, H. (1979). The career of continental drift theory. *Studies in History and Philosophy of Science, 10*, 25.
Franklin, A.D. (1981). Millikan's published and unpublished data on oil drops. *Historical Studies in the Physical Sciences, 11*, 185–201.
Franklin, A.D. (1986). *The Neglect of Experiment*. Cambridge: Cambridge University Press.
Franklin, A.D. (1997). Are there really electrons? Experiment and reality. *Physics Today, 50*, 26–33.
Fuller, S. (2000). *Thomas Kuhn: A Philosophical History of Our Times*. Chicago, IL: University of Chicago Press.
Gabel, D.L. & Bunce, D.M. (1994). Research on problem solving: Chemistry. In D.L. Gabel (ed.), *Handbook of Research on Science Teaching* (pp. 301–326). New York: Macmillan.
Galison, P. (2008). Ten problems in history and philosophy of science. *Isis, 99*, 111–124.
Garber, E., Brush, S.G. & Everitt, C.W.F. (eds) (1986). *Maxwell on Molecules and Gases*. Cambridge, MA: MIT Press.
Gardner, M. (1982). Predicting novel facts. *British Journal for the Philosophy of Science, 33*, 9.
Garfinkel, H. (1967). *Studies in Ethnomethodology*. Englewood Cliffs, NJ: Prentice-Hall.
Gavroglu, K. (1990). The reaction of the British physicists and chemists to van der Waals' early work and to the law of corresponding states. *Historical Studies in the Physical and Biological Sciences, 20*, 199–237.
Gavroglu, K. (2000). Controversies and the becoming of physical chemistry. In P. Machamer, M. Pera & A. Baltas (eds), *Scientific Controversies: Philosophical and Historical Perspectives* (pp. 177–198). New York: Oxford University Press.
Geiger, H. & Marsden, E. (1909). *On a Diffuse Reflection of the Alpha Particles. Proceedings of the Royal Society* (vol. lxxxii). London: Royal Society.
Giere, R.N. (1983). Testing theoretical hypotheses. In J. Earman (ed.), *Testing Scientific Theories* (pp. 269–298). Minneapolis, MN: University of Minnesota Press.
Giere, R.N. (1988). *Explaining Science: A Cognitive Approach*. Chicago, IL: University of Chicago Press.
Giere, R.N. (1999). *Science Without Laws*. Chicago, IL: University of Chicago Press.
Gillies, D. (1993). *Philosophy of Science in the Twentieth Century: Four Central Themes*. Oxford: Blackwell.
Gillies, D. (2002). Lakatos' criticism of Popper. In G. Kampis, L. Kvasz & M. Stöltzner (eds), *Appraising Lakatos: Mathematics, Methodology and the Man* (pp. 13–22). Dordrecht, The Netherlands: Kluwer.
Goldstein, E. (1880). On the electric discharge in rarefied gases. *Philosophical Magazine, 10*, 173.
Goodstein, D. (2001). In defense of Robert Andrews Millikan. *American Scientist, 89*, 54–60.
Gordin, M.D. (2002). The organic roots of Mendeleev's periodic law. *Historical Studies in the Physical Sciences, 32*, 263–290.
Gordin, M.D. (2004). *A Well Ordered Thing: Dmitrii Mendeleev and the Shadow of the Periodic Table*. New York: Basic Books.
Gower, B. (1997). *Scientific Method: An Historical and Philosophical Introduction*. London: Routledge.
Grünbaum, A. (1963). *Philosophical Problems of Space and Time*. New York: Knopf.
Hadzidaki, P., Kalkanis, G. & Stavrou, D. (2000). Quantum mechanics: A systemic component of the modern physics paradigm. *Physics Education, 35*, 386–392.
Hanson, N.R. (1958). *Patterns of Discovery*. Cambridge: Cambridge University Press.
Heilbron, J.L. (1964). *A history of atomic models from the discovery of the electron to the beginnings of quantum mechanics*. Doctoral dissertation, University of California, Berkeley, CA.
Heilbron, J.L. (1966). The work of H.G.J. Moseley. *Isis, 57*, 336–363.
Heilbron, J.L. (1981). Rutherford-Bohr atom. *American Journal of Physics, 49*, 223–231.
Heilbron, J.L. (1985). Bohr's first theories of the atom. In A.P. French & P.J. Kennedy (eds), *Niels Bohr: A Centenary Volume* (pp. 33–49). Cambridge, MA: Harvard University Press.
Heilbron, J.L. & Kuhn, T. (1969). The genesis of the Bohr atom. *Historical Studies in the Physical Sciences, 1*, 211–290.
Heisenberg, W. (1969). *Der teil und das ganze*. Munich: R. Piper.

Hendry, J. (1980). The development of attitudes to the wave-particle duality of light and quantum theory, 1900–1920. *Annals of Science, 37*, 59–79.

Hentschel, K. (1992). Einstein's attitude towards experiments: Testing relativity theory 1907–1927. *Studies in History and Philosophy of Science, 23*(4), 593–624.

Hentschel, K. (2004). Review of Allan Franklin's *Selectivity and Discord: Two problems of experiment. Philosophy of Science, 71*(4), 607–610.

Herron, J.D. (1977). Rutherford and the nuclear atom. *Journal of Chemical Education, 54*, 499.

Hertz, H. (1883). Uber kathodenstrahlen. *Annalen der Physik und Chemie, 19*, 782.

Hertz, H. (1892). Uben den durchgang der kathodenstrahlen durch dunne metallschichten. *Annalen der Physik und Chemie, 45*, 28.

Hettema, H. (1995). Bohr's theory of the atom 1913–1923: A case study in the progress of scientific research programmes. *Studies in History and Philosophy of Modern Physics, 26B*, 307–323.

Hettema, H. & Kuipers, T.A.F. (1988). The periodic table – its formalization, status, and relation to atomic theory. *Erkenntnis, 28*, 387–408.

Hill, J.W. & Petrucci, R.H. (1999). *General Chemistry: An Integrated Approach* (2nd ed.) Upper Saddle River, NJ: Prentice-Hall.

Hintikka, J. (1999). *Inquiry as Inquiry* (Selected papers, vol. 5). Dordrecht, The Netherlands: Kluwer.

Hintikka, J. (2005). Omitting data – ethical or strategic problem? *Synthese, 145*, 169–176.

Hintikka, J. (2006). Personal communication to the author (email, dated August 15, 2006).

Holton, G. (1952). *Introduction to Concepts and Theories of Physical Science*. Cambridge, MA: Addison-Wesley.

Holton, G. (1969a). Einstein and "crucial" experiments. *American Journal of Physics, 37*(10), 968–982.

Holton, G. (1969b). Einstein, Michelson, and the "crucial" experiment. *Isis, 60*, 133–197.

Holton, G. (1978). Subelectrons, presuppositions, and the Millikan-Ehrenhaft dispute. *Historical Studies in the Physical Sciences, 9*, 161–224.

Holton, G. (1986). *The Advancement of Science and Its Burdens*. Cambridge: Cambridge University Press.

Holton, G. (1988). On the hesitant rise of quantum physics research in the United States. In S. Goldberg & R.H. Stuewer (eds), *The Michelson Era in American Science 1870–1930* (pp. 177–205). New York: American Institute of Physics.

Holton, G. (1992). Ernst Mach and the fortunes of positivism in America. *Isis, 83*, 27–60.

Holton, G. (1993). *Science and Anti-science*. Cambridge, MA: Harvard University Press.

Holton, G. (1998). *The Scientific Imagination*. Cambridge, MA: Harvard University Press.

Holton, G. (1999). R.A. Millikan's struggle with the meaning of Planck's constant. *Physics in Perspective, 1*, 231–237.

Holton, G. (2000). Werner Heisenberg and Albert Einstein. *Physics Today, 53*, 38–42.

Howson, C. (ed.) (1976). *Method and Appraisal in the Physical Sciences: The Critical Background to Modern Sciences, 1800–1905*. Cambridge: Cambridge University Press.

Howson, C. (1990). Essay review: The poverty of historicism. *Studies in History and Philosophy of Science, 21*, 173–179.

Howson, C. & Franklin, A. (1991). Maher, Mendeleev and Bayesianism. *Philosophy of Science, 58*, 574–585.

Hudson, R.G. (2003). Novelty and the 1919 eclipse experiments. *Studies in History and Philosophy of Modern Physics, 34*(1), 107–129.

Hulsizer, R.I. & Lazarus, D. (1977). *The World of Physics*. Menlo Park, CA: Addison-Wesley.

Ihde, A.J. (1969). Foreword. In J.W. van Spronsen (ed.), *The Periodic System of Chemical Elements: A History of the First Hundred Years*. Amsterdam: Elsevier.

Ireson, G. (2000). The quantum understanding of pre-university physics students. *Physics Education, 35*, 15–21.

Irwin, A.R. (2000). Historical case studies: Teaching the nature of science in context. *Science Education, 84*, 5–26.

Jaki, S.L. (1984). *Uneasy Genius: The Life and Work of Pierre Duhem*. Dordrecht, The Netherlands: Martinus Nijhoff.

Jammer, M. (1966). *The Conceptual Development of Quantum Mechanics*. New York: McGraw-Hill.

Jaynes, E.T. (1967). Foundations of probability theory and statistical mechanics. *Delaware Seminar in the Foundations of Physics*, *1*, 76–101.

Joesten, M.D., Johnston, D.O., Netterville, J.T. & Wood, J.L. (1991) *World of Chemistry*. Philadelphia, PA: Saunders.

Johnston, I.D., Crawford, K. & Fletcher, P.R. (1998). Student difficulties in learning quantum mechanics. *International Journal of Science Education*, *20*, 427–446.

Jones, L.W. (1977). A review of quark search experiments. *Review of Modern Physics*, *49*, 717–752.

Jones, R.C. (1995). The Millikan oil-drop experiment: Making it worthwhile. *American Journal of Physics*, *63*, 970–977.

Justi, R.S. & Gilbert, J.K. (2000). History and philosophy of science through models: Some challenges in the case of "the atom." *International Journal of Science Education*, *22*, 993–1009.

Kadvany, J. (2001). *Imre Lakatos and the Guises of Reason*. Durham: Duke University Press.

Kaji, M. (2003). Mendeleev's discovery of the periodic law: The origin and the reception. *Foundations of Chemistry*, *5*, 189–214.

Kalkanis, G., Hadzidaki, P. & Stavrou, D. (2003). An instructional model for a radical conceptual change towards quantum mechanics concepts. *Science Education*, *87*, 257–280.

Kapusta, J.I (1975). Best measuring time for a Millikan oil drop experiment. *American Journal of Physics*, *43*, 799–800.

Kaufmann, W. (1897). Die magnetische ablenkbarkeit der kathodenstrahlen und ihre abhängigkeit vom entladungspotential. *Annalen der Physik und Chemie*, *61*, 544.

Khishfe, R. & Lederman, N.G. (2006). Teaching nature of science within a controversial topic: Integrated versus nonintegrated. *Journal of Research in Science Teaching*, *43*, 395–418.

Kitchener, R.F. (1993). Piaget's epistemic subject and science education: Epistemological versus psychological issues. *Science & Education*, *2*, 137–148.

Kitcher, P. (1993). *The Advancement of Science: Science Without Legend, Objectivity Without Illusions*. Oxford: Oxford University Press.

Kitcher, P. (2000). Patterns of scientific controversies. In P. Machamer, M. Pera & A. Baltas (eds), *Scientific Reasoning: Philosophical and Historical Perspectives* (pp. 21–39). New York: Oxford University Press.

Klassen, S. (2009). Identifying and addressing student difficulties with the Millikan oil drop experiment. *Science & Education*, *17* (in press).

Klein, M.J. (1963), Einstein's first paper on quanta. *The Natural Philosopher*, *2*, 59–86.

Klein, M.J. (1964). Einstein and wave–particle duality. *The Natural Philosopher*, *3*, 3–49.

Knight, D. (1998). Preface. In D. Knight & H. Kragh (eds), *The Making of the Chemist: The Social History of Chemistry in Europe, 1789–1914*. Cambridge: Cambridge University Press.

Kohler, R.E. (1971). The origin of Lewis's theory of the shared pair bond. *Historical Studies in the Physical Sciences*, *3*, 343–376.

Kotz, J.C. & Purcell, K.F. (1991). *Chemistry and Chemical Reactivity* (2nd ed.). Philadelphia, PA: Saunders.

Kragh, H. (1999). *Quantum Generations: A History of Physics in the Twentieth Century*. Princeton, NJ: Princeton University Press.

Kruglak, H. (1972). Another look at the Pasco-Millikan oil-drop apparatus. *American Journal of Physics*, *40*, 768–769.

Kubli, F. (1970). *Archives of the history of the exact sciences*, *7*, 26–68.

Kuhn, T.S. (1962). *The Structure of Scientific Revolutions*. Chicago, IL: University of Chicago Press.

Kuhn, T.S. (1970a). *The Structure of Scientific Revolutions* (2nd ed.). Chicago, IL: University of Chicago Press.

Kuhn, T.S. (1970b). Logic of discovery or psychology of research? In I. Lakatos & A. Musgrave (eds), *Criticism and the Growth of Knowledge* (pp. 1–23). Cambridge: Cambridge University Press.

Kuhn, T.S. (1977). The function of measurement in modern physical science. In T.S. Kuhn (ed.) *Essential Tension* (pp. 178–224). Chicago, IL: University of Chicago Press. (originally published in *Isis*, *52*, 161–190, 1961).

Kuhn, T.S. (1978). *Black-Body Theory and the Quantum Discontinuity: 1894–1912*. New York: Oxford University Press.

Kuhn, T.S. (1984). Revisiting Planck. *Historical Studies in the Physical Sciences*, *14*, 231–252.

Lakatos, I. (1970). Falsification and the methodology of scientific research programmes. In I. Lakatos & A. Musgrave (eds), *Criticism and the Growth of Knowledge* (pp. 91–195). Cambridge: Cambridge University Press.

Lakatos, I. (1971). History of science and its rational reconstructions. In R.C. Buck & R.S. Cohen (eds), *Boston Studies in the Philosophy of Science* (vol. 8, pp. 91–136). Dordrecht: Reidel.

Lakatos, I. (1974). The role of crucial experiments. *Studies in History and Philosophy of Science*, *4*, 309–325.

Lakatos, I. (1976). *Proofs and Refutations*. Cambridge: Cambridge University Press.

Lakatos, I. (1999). Lectures on scientific method. In M. Motterlini (ed.), *For and Against Method: Including Lakatos's Lectures on Scientific Method and the Lakatos-Feyerabend Correspondence* (pp. 19–112). Chicago, IL: University of Chicago Press.

Lakatos, I. & Zahar, E. (1978). Why did Copernicus's programme supersede Ptolemy's? In J. Worrall & G. Currie (eds), *The Methodology of Scientific Research Programmes* (philosophical papers, vol. 1, pp. 168–191). Cambridge: Cambridge University Press (first published 1976).

Laloë, F. (2001). Do we really understand quantum mechanics? Strange correlations, paradoxes, and theorems. *American Journal of Physics*, *69*, 655–701.

LaRue, G.S., Phillips, J.D. & Fairbank, W.M. (1981). Observation of fractional charge of (1/3) e on matter. *Physics Review Letters*, *46*, 967.

Laudan, L. (1977). *Progress and Its Problems*. Berkeley, CA: University of California Press.

Laudan, L. (1990). *Science and Relativism*. Chicago, IL: University of Chicago Press.

Laudan, L. (1996). *Beyond Positivism and Relativism: Theory, Method and Evidence*. Boulder, CO: Westview Press.

Laudan, L., Donovan, A., Laudan, R., Barker, P., Brown, H., Leplin, J., Thagard, P. & Wykstra, S. (1986). Scientific change: Philosophical models and historical research. *Synthese*, *69*, 141–223.

Laudan, R., Laudan, L. & Donovan, A. (1988). Testing theories of scientific change. In A. Donovan, L. Laudan & R. Laudan (eds), *Scrutinizing Science: Empirical Studies of Scientific Change* (pp. 3–44). Dordrecht, The Netherlands: Kluwer.

Laymon, R. (1988). The Michelson-Morley experiment and the appraisal of theories. In A. Donovan, L. Laudan & R. Laudan (eds), *Scrutinizing Science: Empirical Studies of Scientific Change* (pp. 245–266). Dordrecht, The Netherlands: Kluwer.

Leary, J.J. & Kippeny, T.C. (1999). A framework for presenting the modern atom. *Journal of Chemical Education*, *76*(9), 1217–1218.

Lederman, N.G. (1992). Students' and teachers' conceptions about the nature of science: A review of the research. *Journal of Research in Science Teaching*, *29*, 331–359.

Lederman, N.G. (2004). Syntax of nature of science within inquiry and science instruction. In L.B. Flick & N.G. Lederman (eds), *Scientific Inquiry and Nature of Science* (pp. 301–317). Dordrecht, The Netherlands: Kluwer.

Lenard, P.E.A. (1902). über die LichtelektrischeWirkung. *Annalen der Physik*, *8*, 149–198.

Lewin, K. (1935). The conflict between Aristotelian and Galilean modes of thought in contemporary psychology. In K. Levin (ed.) *A dynamic theory of personality* (pp. 1–42). New York: McGraw-Hill (first published 1931, *Journal of Genetic Psychology*).

Lewis, G.N. (1913). Valence and tautomerism. *Journal of American Chemical Society*, *35*, 1448–1455.

Lewis, G.N. (1916). The atom and the molecule. *Journal of American Chemical Society*, *38*, 762–785.

Lewis, G.N. (1919). Letter to Robert Millikan (reproduced in Kohler, 1971).

Lewis, G.N. (1923) *Valence and the Structure of Atoms and Molecules*. New York: Chemical Catalog.

Lin, H. & Chen, C. (2002). Promoting preservice chemistry teachers' understanding about the nature of science through history. *Journal of Research in Science Teaching*, *39*, 773–792.

Lindemann, F.A. (1913–1914). Atomic models and X-ray spectra. *Nature*, *92*, 500–501.

Lippincott, W.T., Garrett, A.B. & Verhoek, F.H. (1977). *Chemistry: A Study of Matter* (3rd ed.). New York: Wiley.

Lipton, P. (1991). *Inference to the Best Explanation*. London: Routledge.

Lipton, P. (2005a). Testing hypotheses: Prediction and prejudice. *Science, 307*(14 January), 219–221.

Lipton, P. (2005b). Response. *Science, 308*(3 June), 1411–1412.

Lockyer, N. (1881). Solar physics – the chemistry of the sun. *Nature, 24*, 267.

Lockyer, N. (1897). On the chemistry of the hottest stars. *Proceedings of the Royal Society, 61*, 148.

Lorentz, H.A. (1895). *Versuch einer theorie de electrischen und optischen erscheinungen in bewegten Körpern* (collected papers, vol. 5). Leiden: Brill.

Lorentz, H.A. (1952). *The Theory of Electrons*. New York: Dover (original work Published 1909).

Losee, J. (2001). *A Historical Introduction to the Philosophy of Science* (4th ed.). Oxford: Oxford University Press.

Machamer, P., Pera, M. & Baltas, A. (2000). Scientific controversies: An introduction. In P. Machamer, M. Pera & A. Baltas (eds), *Scientific Controversies: Philosophical and Historical Perspectives* (pp. 3–17). New York: Oxford University Press.

Mahan, B. & Myers, R.J. (1987). *University Chemistry* (4th ed., Spanish). Menlo Park, CA: Benjamin Cummings.

Maher, P. (1988). Prediction accommodation and the logic of discovery. In A. Fine & J. Leplin (eds), *PSA 1988* (vol. 1). East Lansing: Philosophy of Science Association.

Mannheim, K. (1952). *Essays on the Sociology of Knowledge*. London: Routledge & Kegan Paul.

Mar, N.M., Lee, E.R., Fleming, G.R., Casey, B.C.K., Perl, M.L., Garwin, E.L., Hendricks, C.D., Lackner, K.S. & Shaw, G.L. (1996). Improved search for elementary particles with fractional electric charge. *Physical Review D, 53*, 6017–6032.

Margenau, H. (1950). *The Nature of Physical Reality*. New York: McGraw-Hill.

Martin, R.N.D. (1991). *Pierre Duhem: Philosophy and History in the Work of a Believing Physicist*. La Salle, IL: Open Court.

Matthews, M.R. (1987). *Experiment as the Objectification of Theory: Galileo's Revolution. Proceedings of the Second International Seminar on Misconceptions and Educational Strategies in Science and Mathematics* (vol. 1, pp. 289–298). Ithaca, NY: Cornell University.

Matthews, M.R. (1994). *Science Teaching: The Role of History and Philosophy of Science*. New York: Routledge.

Maxwell, J.C. (1860). Illustrations of the dynamical theory of gases. *Philosophical Magazine, 19*, 19–32 (reproduced in *Scientific Papers*, 1965 (pp. 377–409). New York: Dover).

Maxwell, J.C. (1874a). Plateau on soap bubbles. *Nature, 10*, 119–121.

Maxwell, J.C. (1874b). Van der Waals on the continuity of the gaseous and liquid states. *Nature, 10*, 471–478.

Maxwell, J.C. (1875). On the dynamical evidence of the molecular constitution of bodies. *Journal of the Chemical Society, 13*, 493–508 (reproduced in *Scientific Papers*, 1965 (pp. 418–438)).

Maxwell, J.C. (1965). *Scientific Papers*. New York: Dover.

Mayo, D. (1991). Novel evidence and severe tests. *Philosophy of Science, 58*, 523–552.

Mayo, D. (1996). *Error and the Growth of Experimental Knowledge*. Chicago, IL: University of Chicago Press.

Mayo, D. (2003). Novel work on problems of novelty? Comments on Hudson. *Studies in History and Philosophy of Modern Physics, 34*(1), 131–134.

McComas, W.F., Almazroa, H. & Clough, M.P. (1998). The role and character of the nature of science in science education. *Science & Education, 7*, 511–532.

McMullin, E. (1985). Galilean idealization. *Studies in History and Philosophy of Science, 16*, 247–273.

McMullin, E. (1997). Review of P. Duhem. Essays in the history and philosophy of science, Hackett, 1996. *British Journal for the Philosophy of Science, 48*(4), 606–609.

McMurry, J. & Fay, R.C. (2001). *Chemistry* (3rd ed.) Upper Saddle River, NJ: Prentice-Hall.

Medicus, H.A. (1974). Fifty years of matter waves. *Physics Today, 27*, 38–45.

Mellado, V., Ruiz, C., Bermejo, M.L. & Jiménez, R. (2006). Contributions from the philosophy of science to the education of science teachers. *Science & Education, 15*, 419–445.

Mendeleev, D. (1869). Ueber die beziehungen der eigenschaften zu den atom gewichten der elemente. *Zeitschrift für Chemie*, *12*, 405–406 (English trans. by C. Giunta).

Mendeleev, D. (1879). The periodic law of the chemical elements. *The Chemical News*, *40*(1042), 243–244.

Mendeleev, D. (1889). The periodic law of the chemical elements. *Journal of the Chemical Society*, *55*, 634–656 (Faraday lecture, delivered on 4 June 1889).

Mendeleev, D. (1897). *The Principles of Chemistry* (2nd English ed., trans. of 6th Russian ed.). New York: American Home Library Company.

Mendoza, E. (1975). A critical examination of Herapath's dynamical theory of gases. *British Journal for the History of Science*, *8*, 155–165.

Merton, R.K., Sills, D.L. & Stigler, S.M. (1984). The Kelvin dictum and social science: An excursion into the history of an idea. *Journal of the History of the Behavioral Sciences*, *20*, 319–331.

Merz, J.T. (1904). *A History of European Thought in the Nineteenth Century* (vol. 1). London: William Blackwood.

Michelini, M., Ragazzon, R., Santi, L. & Stefanel, A. (2000). Proposal for quantum physics in secondary school. *Physics Education*, *35*, 406–410.

Michell, J. (2003). The quantitative imperative: Positivism, naïve realism and the place of qualitative methods in psychology. *Theory & Psychology*, *13*, 5–31.

Michelson, A.A. (1927). *Studies in Optics*. Chicago, IL: University of Chicago Press.

Michelson, A.A. & Morley, E.W. (1887). On the relative motion of the earth and the luminiferous ether. *American Journal of Science*, *34*(3rd Series), 333–345.

Millikan, R.A. (1910). A new modification of the cloud method of determining the elementary electrical charge and the most probable value of that charge. *Philosophical Magazine*, *19*, 209–228.

Millikan, R.A. (1911). The isolation of an ion, a precision measurement of its charge, and the correction of Stokes's law. *Physical Review*, *32*, 349.

Millikan, R.A. (1913a). On the cause of the apparent differences between spark and arc sources in the imparting of initial speeds to Photo-Electrons. *Physical Review*, *1*, 73–75.

Millikan, R.A. (1913b). On the elementary electrical charge and the Avogadro constant. *Physical Review*, *2*, 109–143.

Millikan, R.A. (1914). A direct determination of "*h*." *Physical Review*, *4*, 73–75.

Millikan, R.A. (1915). New Tests of Einstein's photo-electric equation. *Physical Review*, *6*, 55.

Millikan, R.A. (1916a), Einstein's photoelectric equation and contact electromotive force. *Physical Review*, *7*, 18–32.

Millikan, R.A. (1916b). A direct photoelectric determination of Planck's "*h*." *Physical Review*, *7*, 355–388.

Millikan, R.A. (1916c). The existence of a subelectron? *Physical Review*, *8*, 595–625.

Millikan, R.A. (1917). *The Electron*. Chicago, IL: University of Chicago Press.

Millikan, R.A. (1947). *Electrons (+ and −), Protons, Photons, Neutrons, Mesotrons, and Cosmic Rays* (2nd ed.), Chicago, IL: University of Chicago Press (first published 1935).

Millikan, R.A. (1949). Albert Einstein on his seventieth birthday. *Reviews of Modern Physics*, *21*, 343–344.

Millikan, R.A. (1950). *The Autobiography of Robert A. Millikan*. New York: Prentice-Hall.

Millikan, R.A. (1965). *The electron and the light-quant from the experimental point of view (Nobel prize acceptance speech, 1923). Nobel Lectures: Physics* (pp. 54–66). Amsterdam: Elsevier.

Millikan, R.A. & Begeman, L. (1908). On the charge carried by the negative ion of an ionized gas. *Physical Review*, *26*, 197–198.

Monk, M. & Osborne, J. (1997). Placing the history and philosophy of science on the curriculum: A model for the development of pedagogy. *Science Education*, *81*, 405–424.

Mortimer, C.E. (1983) *Chemistry* (5th ed.). Belmont, CA: Wadsworth.

Moseley, H.G.J. (1913). High frequency spectra of the elements. *Philosophical Magazine*, *26*, 1025–1034.

Moseley, H.G.J. (1913–1914). Atomic models and X-ray spectra. *Nature*, *92*, 554.

Moseley, H.G.J. (1914). High frequency spectra of the elements, Part II. *Philosophical Magazine, 27*, 703–713.

Motterlini, M. (1999). Introduction: A dialogue. In M. Motterlini (ed.), *For and Against Method: Including Lakatos's Lectures on Scientific Method and the Lakatos-Feyerabend Correspondence* (pp. 1–18). Chicago, IL: University of Chicago Press.

Motterlini, M. (2002). Reconstructing Lakatos: A reassessment of Lakatos's epistemological project in the light of the Lakatos Archive. *Studies in History and Philosophy of Science, 33*(3), 487–509.

Moyer, D. (1979). Revolution in science: The 1919 eclipse test of general relativity. In A. Perlmutter & L. Scott (eds), *On the Path of Albert Einstein* (pp. 55–102). New York: Plenum.

Murphy, N. (1989). Another look at novel facts. *Studies in History and Philosophy of Science, 20*, 385–388.

Musgrave, A. (1974). Logical vs. historical theories of confirmation. *British Journal for the Philosophy of Science, 25*, 16.

Nagaoka, H. (1904). Kinetics of system of particles illustrating the band and the phenomena of radioactivity. *Philosophical Magazine, 7*, 445–455.

Needham, P. (1998). Duhem's physicalism. *Studies in History and Philosophy of Science, 29*(1), 33–62.

Niaz, M. (1993). If Piaget's epistemic subject is dead, shall we bury the scientific research methodology of idealization? *Journal of Research in Science Teaching, 30*, 809–812.

Niaz, M. (1994). Enhancing thinking skills: Domain specific/domain general strategies—A dilemma for science education. *Instructional Science, 22*, 413–422.

Niaz, M. (1995). Progressive transitions from algorithmic to conceptual understanding in student ability to solve chemistry problems: A Lakatosian interpretation. *Science Education, 79*, 19–36.

Niaz, M. (1998). From cathode rays to alpha particles to quantum of action: A rational reconstruction of structure of the atom and its implications for chemistry textbooks. *Science Education, 82*, 527–552.

Niaz, M. (1999a). The role of idealization in science and its implications for science education. *Journal of Science Education and Technology, 8*, 145–150.

Niaz, M. (1999b). Should we put observations first? *Journal of Chemical Education, 75*, 734.

Niaz, M. (2000a). A rational reconstruction of the kinetic molecular theory of gases based on history and philosophy of science and its implications for chemistry textbooks. *Instructional Science, 28*, 23–50.

Niaz, M. (2000b). The oil drop experiment: A rational reconstruction of the Millikan-Ehrenhaft controversy and its implications for chemistry textbooks. *Journal of Research in Science Teaching, 37*, 480–508.

Niaz, M. (2001a). Understanding nature of science as progressive transitions in heuristic principles. *Science Education, 85*, 684–690.

Niaz, M. (2001b). A rational reconstruction of the origin of the covalent bond and its implications for general chemistry textbooks. *International Journal of Science Education, 23*(6), 623–641.

Niaz, M. (2005a). The quantitative imperative vs. The imperative of presuppositions. *Theory & Psychology, 15*, 247–256

Niaz, M. (2005b). An appraisal of the controversial nature of the oil drop experiment: Is closure possible? *British Journal for the Philosophy of Science, 56*(4), 681–702.

Niaz, M. (2008a). Do we need to write physical science textbooks within a history and philosophy of science perspective. In M.V. Thomase (ed.), *Science Education in Focus* (pp. 55–105). New York: Nova Science.

Niaz, M. (2009). Progressive transitions in chemistry teachers' understanding of nature of science based on historical controversies. *Science & Education, 18*(1) pp. 43–65.

Niaz, M. (2008c). *Teaching General Chemistry: A History and Philosophy of Science Approach.* New York: Nova Science.

Niaz, M., Aguilera, D., Maza, A. & Liendo, G. (2002). Arguments, contradictions, resistances, and conceptual change in students' understanding of atomic structure. *Science Education, 86*, 505–525.

Niaz, M. & Fernández, R. (2008). Understanding quantum numbers in general chemistry textbooks. *International Journal of Science Education, 30*(7), 869–901.

Niaz, M., Klassen, S., McMillan, B. & Metz, D. (2009). Leon Cooper's perspective on teaching science: An interview study. *Science & Education, 17* (in press).

Niaz, M. & Robinson, W.R. (1992). From "algorithmic mode" to "conceptual gestalt" in understanding the behavior of gases: An epistemological perspective. *Research in Science & Technological Education, 10*(1), 53–64.

Niaz, M., Rodríguez, M.A. & Brito, A. (2004). An appraisal of Mendeleev's contribution to the development of the periodic table. *Studies in History and Philosophy of Science, 35*, 271–282.

Nicholson, J.W. (1911). A structural theory of the chemical elements. *Philosophical Magazine, 22*, 864–889.

Nicholson, J.W. (1912). The constitution of the solar corona II. *Monthly Notices of the Royal Astronomical Society, 72*, 677–692.

Noyes, P. (1984). The downfall of determinism (book review). *Physics Today, 37*(10), 95.

Nunan, R. (1984). Novel facts, Bayesian rationality, and the history of continental drift. *Studies in History and Philosophy of Science, 15*, 279.

Nyhof, J. (1988). Philosophical objections to the kinetic theory. *British Journal for the Philosophy of Science, 39*, 81–109.

Ohanian, H.C. (1987). *Modern Physics*. Englewood Cliffs, NJ: Prentice-Hall.

Olenick, R.P., Apostol, T.M. & Goodstein, D.L. (1985). *Beyond the Mechanical Universe. From Electricity to Modern Physics*. New York: Cambridge University Press.

Ostwald, W. (1927). *Lebenslinien, 2*, 178–179.

Olwell, R. (1999). Physical isolation and marginalization in physics: David Bohm's cold war exile. *Isis, 90*, 738–756.

Pais, A. (1982). *"Subtle Is the Lord ..." The Science and the Life of Albert Einstein*. Oxford: Oxford University Press.

Papineau, D. (1979). *Theory and Meaning*. Oxford: Clarendon.

Parson, A.L. (1915). A magneton theory of the structure of the atom. *Smithsonian Miscellaneous Collections, 65*, 1–80.

Pascual-Leone, J. (1978). Compounds, confounds, and models in developmental information processing: A reply to Trabasso and Foellinger. *Journal of Experimental Child Psychology, 26*, 18–40.

Pauli, W. (1925). Über den zusammenhang des abschlußes der elektronengruppen im atom mit der komplexstruktur der spektren. *Zeitschrift für Physik, 31*, 765–785.

Pearson, K. (1951). *The Grammar of Science*. London: J.M. Dent (first published in 1892).

Perl, M.L. (2005). Personal communication to the author (email dated, December 1).

Perl, M.L. & Lee, E.R. (1997). The search for elementary particles with fractional electric charge and the philosophy of speculative experiments. *American Journal of Physics, 65*, 698–706.

Perl, M.L., Lee, E.R. & Loomba, D. (2004). A brief review of the search for isolatable fractional charge elementary particles. *Modern Physics Letters, A19*, 2595–2610.

Perrin, C.E. (1988). The chemical revolution. In A. Donovan, L. Laudan & R. Laudan (eds), *Scrutinizing Science: Empirical Studies of Scientific Change* (pp. 105–124). Dordrecht, The Netherlands: Kluwer.

Petzoldt, J. (1913). Positivistische philosophie. *Zeitschrift für Positivistische Philosophie, 1*, 3–4.

Phillips, J.S., Strozak, V.S. & Wistrom, C. (2000). *Chemistry: Concepts and Applications* (Spanish ed.). New York: McGraw-Hill.

Pickering, A. (1984). *Constructing Quarks: A Sociological History of Particle Physics*. Chicago, IL: University of Chicago Press.

Pickering, M. (1990). Further studies on concept learning versus problem solving. *Journal of Chemical Education, 67*, 254–255.

Pocoví, M.C. (2007). The effects of a history based instructional material on the students' understanding of field lines. *Journal of Research in Science Teaching, 44*, 107–132.

Polanyi, M. (1964). *Personal Knowledge*. Chicago, IL: University of Chicago Press (first published 1958).

Pomeroy, D. (1993). Implications of teachers' beliefs about the nature of science. *Science Education, 77*, 261–278.

Popper, K. (1959). *The Logic of Scientific Discovery*. New York: Basic Books.

Popper, K. (1965). *Conjectures and Refutations: The Growth of Scientific Knowledge*. New York: Harper & Row.

Pospiech, G. (2000). Uncertainty and complementarity: The heart of quantum physics. *Physics Education, 35*, 393–399.

Porter, T.M. (1981). A statistical survey of gases: Maxwell's social physics. *Historical Studies in Physical Sciences, 12*, 77–116.

Porter, T.M. (1984). From Quetelet to Maxwell: Social statistics and the origins of statistical physics. In I.B. Cohen (ed.), *Boston Studies in the Philosophy of Science* (vol. 150, pp. 345–362). Dordrecht, The Netherlands: Kluwer.

Przibram, K. (1907). *Physikalische Zeitschrift, 8*, 489–491.

Quagliano, J.V. & Vallarino, L.M. (1969). *Chemistry* (3rd ed.) Englewood Cliffs, NJ: Prentice-Hall.

Quine, W.V.O. (1953). *From a Logical Point of View*. New York: Harper and Row.

Quine, W.V.O. (1969). Epistemology naturalized. In W.V.O. Quine (ed.) *Ontological Relativity and other Essays* (pp. 69–90). New York: Columbia University Press.

Raman, V.V. & Forman, P. (1969). Why was it Schrödinger who developed de Broglie's ideas? *Historical Studies in the Physical Sciences, 1*, 291–314.

Reichenbach, H. (1942). *From Copernicus to Einstein*. New York: Philosophical Library.

Rigden, J.S. & Stuewer, R.H. (2005). Do physicists understand physics? *Physics in Perspective, 7*, 387–389.

Rodebush, W.H. (1928). The electron theory of valence. *Chemical Review, 5*, 509–531.

Rodríguez, M.A. & Niaz, M. (2004a). A reconstruction of structure of the atom and its implications for general physics textbooks. *Journal of Science Education and Technology, 13*, 409–424.

Rodríguez, M.A. & Niaz, M. (2004b). The oil drop experiment: An illustration of scientific research methodology and its implications for physics textbooks. *Instructional Science, 32*, 357–386.

Rodríguez, M.A. & Niaz, M. (2004c) La teoría cinético-molecular de los gases en libros de física: Una perspectiva basada en la historia y filosofía de la ciencia. *Revista de Educación en Ciencias, 5*, 68–72.

Rosenfeld, L. (1963). *Introduction to Bohr's on the Constitution of Atoms and Molecules* (pp. xliv–xlv). Copenhagen: Munksgaard.

Rosnick, P. & Clement, J. (1980). Learning without understanding: The effect of tutoring strategies on algebra misconceptions. *Journal of Mathematical Behavior, 3*, 3–28.

Russell, H.N. (1920). Note on the Sobral eclipse photographs. *Monthly Notices of the Royal Astronomical Society, 81*, 154–164.

Rutherford, E. (1904). *Radioactivity*. Cambridge: Cambridge University Press.

Rutherford, E. (1906). *Radioactive Transformations*. London: Constable.

Rutherford, E. (1909) Address to the mathematical and physical science section, Report of the 79 Meeting of the British Association for the Advancement of science, Winnipeg, Canada, Aug. 25–Sept. 1, London, Murray, pp. 373–385.

Rutherford, E. (1911). The scattering of alpha and beta particles by matter and the structure of the atom. *Philosophical Magazine, 21*, 669–688.

Rutherford, E. (1913). *Radioactive Substances and Their Radiations*. Cambridge: Cambridge University Press.

Rutherford, E. (1915). The constitution of matter and the evolution of the elements. In *Address to the annual meeting of the National Academy of Sciences* (pp. 167–202). Washington, DC: Smithsonian Institution.

Rutherford, E. & Geiger, H. (1908). The charge and the nature of the α-particle. *Proceedings of the Royal Society (London), 81*, 168–171.

Sachs, M. (1988). *Einstein Versus Bohr: The Continuing Controversies in Physics*. La Salle, IL: Open court.

Schrödinger, (1922). Bemerkenswerte eigenschaft der quantenbahnen? *Zeitschrift fur Physik, 12*, 13–23.

Schrödinger, E. (1926a). On Einstein's gas theory. *Physikalische Zeitschrift, 27*, 95.
Schrödinger, E. (1926b). *Annalen de Physik, 79*, 734.
Schuster, A. (1890). The discharge of electricity through gases. *Proceedings of the Royal Society A, 47*, 526.
Schwab, J.J. (1962). *The Teaching of Science as Enquiry*. Cambridge, MA: Harvard University Press.
Schwab, J.J. (1974). The concept of the structure of a discipline. In E.W. Eisner & E. Vallance (eds), *Conflicting Conceptions of Curriculum*. Berkeley, CA: McCutchan (first published in 1962).
Sears, F.W., Zemansky, M.W. & Young, H.D. (1991). *University Physics*. Reading, MA: Addison-Wesley.
Segal, B.G. (1989). *Chemistry: Experiment and Theory* (2nd ed.). New York: Wiley.
Segerstrale, U. (1995). Good to the last drop? Millikan stories as "canned" pedagogy. *Science and Engineering Ethics, 1*(3), 197–214.
Segre, M. (1989). Galileo, Viviani, and the tower of Pisa. *Studies in History and Philosophy of Science, 20*, 435–451.
Serway, R.A. (1996). *Physics for Scientists and Engineers with Modern Physics*. New York: McGraw-Hill.
Segal, B.G. (1989). *Chemistry: Experiment and Theory* (2nd ed.). New York: Wiley.
Shankland, R.S. (1963). Conversations with Albert Einstein. *American Journal of Physics, 31*, 47–57.
Shankland, R.S. (1964). The Michelson-Morley experiment. *American Journal of Physics, 32*, 23.
Shapere, D. (1977). Scientific theories and their domains. In F. Suppe (ed.), *The Structure of Scientific Theories* (2nd ed., pp. 518–565). Chicago, IL: University of Illinois Press.
Shiland, T.W. (1995). What's the use of all this theory? The role of quantum mechanics in high school chemistry textbooks. *Journal of Chemical Education, 72*, 215–219.
Shiland, T.W. (1997). Quantum mechanics and conceptual change in high school chemistry textbooks. *Journal of Research in Science Teaching, 34*, 535–545.
Shimony, A. (1985). Review of Folse's "The philosophy of Niels Bohr." *Physics Today, 38*, 109.
Sisler, H.H., Dresdner, R.D. & Mooney, W.T. (1980). *Chemistry: A Systematic Approach*. New York: Oxford University Press.
Smith, M.U. & Scharmann, L.C. (1999). Defining versus describing the nature of science: A pragmatic analysis for classroom teachers and science educators. *Science Education, 83*, 493–509.
Snow, C.P. (1981). *The Physicists*. Boston, MA: Little, Brown.
Sponsel, A. (2002). Constructing a "revolution in science": The campaign to promote a favourable reception for the 1919 solar eclipse experiments. *British Journal for the History of Science, 35*(4), 439–467.
Stanley, M. (2003). An expedition to heal the wounds of war: The 1919 eclipse and Eddington as Quaker adventurer. *Isis, 94*, 57–89.
Stinner, A. (1989). The teaching of physics and the contexts of inquiry: From Aristotle to Einstein. *Science Education, 73*, 591–605.
Stinner, A. (1992). Science textbooks and science teaching: From logic to evidence. *Science Education, 76*, 1–16.
Stinner, A. (2000). Review of *Isaac Newton: Adventurer in thought* by A. Rupert Hall. *Science Education, 84*, 280–286.
Stinner, A. & Metz, D. (2006). Thought experiments, Einstein, and physics education. *Physics in Canada* (November/December), 361–372.
Stoker, H.S. (1990). *Introduction to Chemical Principles* (3rd ed.). New York: Macmillan.
Stuewer, R.H. (1970). Non-Einsteinian Interpretations of the Photoelectric Effect. In R.H. Stuewer (ed.), *Historical and Philosophical Perspectives of Science* (pp. 246–263). Minneapolis, MN: University of Minnesota Press.
Stuewer, R.H. (1975). *The Compton Effect: Turning Point in Physics*. New York: Science History Publications.
Styer, D.F. (2000). *The Strange World of Quantum Mechanics*. Cambridge: Cambridge University Press.

Suvorov, S.G. (1966). Einstein's philosophical views and their relation to his physical opinions. *Soviet Physics Ospekhi*, *8*, 578.

Taber, K.S. (2005). Learning quanta: Barriers to stimulating transitions in student understanding of orbital ideas. *Science Education*, *89*, 94–116.

Thomason, N. (1992). Could Lakatos, even with Zahar's criterion for novel fact, evaluate the Copernican research programme? *British Journal for the Philosophy of Science*, *43*, 161–200.

Thomson, G.P. & Reid, A. (1928). *Proceedings of the Royal Society of London*, A, *117*, 601–609.

Thomson, J.J. (1881). On the electric and magnetic effects produced by the motion of electrified bodies. *Philosophical Magazine*, *11*, 229.

Thomson, J.J. (1897). Cathode rays. *Philosophical Magazine*, *44*, 293–316.

Thomson, J.J. (1898). *Philosophical Magazine*, *46*, 528.

Thomson, J.J. (1907). *The Corpuscular Theory of Matter*. London: Constable.

Thomson, J.J. (1909). *Proceedings of the Cambridge Literary and Philosophical Society*, *15* (part 5).

Thomson, J.J. (1914). The forces between atoms and chemical affinity. *Philosophical Magazine*, *27*, 757–789.

Thomson, J.J. (Chair). (1919). Joint eclipse meeting of the Royal Society and the Royal Astronomical Society. *Observatory*, *42*, 389–398.

Thomson, J.J. (1925. *Structure of Light*. Cambridge: Cambridge University Press.

Thomson, W. (1891). *Popular Lectures and Addresses* (vol. 1). London: Macmillan.

Townsend, J.S. (1897). *Proceedings of the Cambridge Philosophical Society*, *9*, 244.

Tsaparlis, G. (1997). Atomic orbitals, molecular orbitals and related concepts: Conceptual difficulties among chemistry students. *Research in Science Education*, *27*, 271–287.

Tsaparlis, G. (2001). Towards a meaningful introduction to the Schrödinger equation through historical and heuristic approaches. *Chemistry Education: Research and Practice in Europe*, *2*, 203–213.

Umland, J.B. (1993). *General Chemistry*. St. Paul, MN: West Publishing.

Urbach, P. (1989). The Bayesian alternative to the methodology of scientific research programmes. In K. Gavroglu, Y. Goudaroulis & P. Nicolacopoulos (eds), *Imre Lakatos and Theories of Scientific Change* (pp. 399–412). Dordrecht, The Netherlands: Kluwer.

Van der Waals, J.D. (1873). Over de continuiteit van den gas en vloeistoftoestand. Leyden: University of Leyden.

Van Spronsen, J. (1969). *The Periodic System of Chemical Elements. A History of the First Hundred Years*. Amsterdam: Elsevier.

Vigier, J.-P. (1982). Non-locality, causality, and aether in quantum mechanics. *Astronomische Nachrichten*, *303*, 55–80.

Vihalemm, R. (2003). Are laws of nature and scientific theories peculiar in chemistry? Scrutinizing Mendeleev's discovery. *Foundations of Chemistry*, *5*, 7–22.

Wartofsky, M.W. (1968). *Conceptual Foundations of Scientific Thought: An Introduction to the Philosophy of Science*. New York: Macmillan.

Warwick, A. (1992). Cambridge mathematics and Cavendish physics: Cunningham, Campbell and Einstein's relativity 1905–1911. *Studies in History and Philosophy of Science*, *23*, 625–656.

Weidemann, E. (1884). On the electric discharge in gases. *Philosophical Magazine*, *18*, 35.

Weinert, F. (1995). The Duhem-Quine thesis revisited. *International Studies in the Philosophy of Science*, *9*(2), 147–156.

Wheaton, B.R. (1978). Philip Lenard and the photoelectric effect, 1889–1911. *Historical Studies in the Physical Sciences*, *9*, 299–322.

Wheaton, B.R. (1983). *The Tiger and the Shark: Empirical Roots of Wave-Particle Dualism*. Cambridge: Cambridge University Press.

Wiechert, E. (1897). Ergebniss einer messung der geschwindigkeit der kathodenstrahlen. *Schriften der physikalischökonomisch Gesellschaft zu Königsberg*, *38*, 3.

Will, C.M. (1990). General relativity at 75: How right was Einstein? *Science*, *250*, 770–776.

Wilson, H.A. (1903). Determination of the charge on the ions produced in air by Röntgen rays. *Philosophical Magazine*, *5*, 429–440.

Wilson, D. (1983). *Rutherford: Simple Genius*. Cambridge, MA: MIT Press.

Wittmann, M.C., Steinberg, R.N. & Redish, E.F. (2002). Investigating student understanding of quantum physics: Spontaneous models of conductivity. *American Journal of Physics*, *70*, 218–226.
Worrall, J. (1985). Scientific discovery and theory-confirmation. In J.C. Pitt (ed.), *Change and Progress in Modern Science* (pp. 301–332). Dordrecht: Reidel.
Worrall, J. (1989). Fresnel, Poisson and the white spot. In D. Gooding, Pinch, T. & Schaffer, S. (eds), *The Uses of Experiment* (pp. 135–157). Cambridge: Cambridge University Press.
Zahar, E. (1973). Why did Einstein's programme supersede Lorentz's? *British Journal for the Philosophy of Science*, *24*, 95–123, 223–262.
Ziman, J. (1978). *Reliable Knowledge. An Exploration of the Grounds for Belief in Science*. Cambridge: Cambridge University Press.
Zumdahl, S.S. (1990). *Introductory Chemistry: A Foundation*. Lexington, MA: Heath.
Zumdahl, S.S. (1993). *Chemistry* (3rd ed.) Lexington, MA: Heath.

Author Index

Subject Index

Printed in the United States
141283LV00002B/53/P

9 781402 096259